NanoScience and Technology

NanoScience and Technology

Series Editors:
P. Avouris   B. Bhushan   D. Bimberg   K. von Klitzing   H. Sakaki   R. Wiesendanger

The series NanoScience and Technology is focused on the fascinating nano-world, meso-scopic physics, analysis with atomic resolution, nano and quantum-effect devices, nano-mechanics and atomic-scale processes. All the basic aspects and technology-oriented de-velopments in this emerging discipline are covered by comprehensive and timely books. The series constitutes a survey of the relevant special topics, which are presented by lea-ding experts in the field. These books will appeal to researchers, engineers, and advanced students.

S. Morita   (Ed.)

Roadmap of Scanning Probe Microscopy

With 96 Figures and 6 Tables

 Springer

Professor Dr. Seizo Morita
Graduate School of Engineering, Osaka University
Department of Electrical, Electronic and Information Engineering
Yamada-Oka 2-1, Suita 565-0871, Japan
E-mail: smorita@ele.eng.osaka-u.ac.jp

Series Editors:

Professor Dr. Phaedon Avouris
IBM Research Division
Nanometer Scale Science & Technology
Thomas J. Watson Research Center
P.O. Box 218
Yorktown Heights, NY 10598, USA

Professor Dr. Bharat Bhushan
Ohio State University
Nanotribology Laboratory
for Information Storage
and MEMS/NEMS (NLIM)
Suite 255, Ackerman Road 650
Columbus, Ohio 43210, USA

Professor Dr. Dieter Bimberg
TU Berlin, Fakutät Mathematik/
Naturwissenschaften
Institut für Festkörperphyisk
Hardenbergstr. 36
10623 Berlin, Germany

Professor Dr., Dres. h.c. Klaus von Klitzing
Max-Planck-Institut
für Festkörperforschung
Heisenbergstr. 1
70569 Stuttgart, Germany

Professor Hiroyuki Sakaki
University of Tokyo
Institute of Industrial Science
4-6-1 Komaba, Meguro-ku
Tokyo 153-8505, Japan

Professor Dr. Roland Wiesendanger
Institut für Angewandte Physik
Universität Hamburg
Jungiusstr. 11
20355 Hamburg, Germany

ISSN 1434-4904
ISBN-10 3-540-34314-8 Springer Berlin Heidelberg New York
ISBN-13 978-3-540-34314-1 Springer Berlin Heidelberg New York

Library of Congress Control Number: 2006930878

Springer is a part of Springer Science+Business Media.
springer.com
© Springer-Verlag Berlin Heidelberg 2007

Cover background image: "Single-wall nanotube METFET inverter: organic molecules wrap around a metallic tube channel and modify it to a semiconductor while the second tube is used for gating". Concept: S.V. Rotkin, Physics Department, Lehigh University, image: B. Grosser, Imaging Technology Group, Beckman Institute, UIUC.

Typesetting: SPi, Pondicherry
Cover design: *design& production*, Heidelberg

Printed on acid-free paper   SPIN: 11573982   57/3100/SPi - 5 4 3 2 1 0

Preface

Since the invention of scanning tunneling microscopy (STM) by G. Binnig and H. Rohrer in 1982, various kinds of scanning probe microscopies (SPM) have been developed such as atomic force microscopy (AFM) and scanning near-field optical microscopy (SNOM) in addition to STM. Now STM has plenty of functions such as atomically resolved imaging, scanning tunneling spectroscopy (STS), atom/molecule manipulation, bottom-up nanostructuring, and inelastic electron tunneling spectroscopy (IETS). Thus, STM has achieved remarkable progress, and hence has become a key technology for surface science. Recently, AFM, which is an alternative atomic tool, has also been rapidly developing and achieving remarkable progress such as in true atomic resolution in UHV/gas/liquid, atom/molecule identification, atom/molecule manipulation, bottom-up nanostructuring, and high-speed imaging.

The accelerated development of information and communication technology (ICT) is based on rapidly developing electronics, while the rapid progress and development of electronics is based on the accelerated miniaturization of semiconductor integrated circuits. Therefore, the role of SPM, which is the tool for observation and characterization with the spatial resolution of single nanometer and subnanometer (=atomic scale=angstrom), is becoming increasingly important because of the progress of rapid miniaturization.

"SPM Roadmap" project has been supported by the Japan Society for the Promotion of Science (JSPS) and carried out under the administration of "167th Committee on Nano-Probe Technology of JSPS." "Roadmap of SPM" is the second future prospect on SPM following the previous "Roadmap 2000 of SPM" in Japanese. It should be noted that "Roadmap of SPM" is the first English book that predicts the future development of all the SPM.

In this book, we introduce our future prospect on SPM research and development related to the improvement toward better sensitivity, better performance, greater functionality, better spatial resolution and higher measurement speed. It will enable us to measure various physical properties of diverse materials under wider varieties of environments. Such prediction will determine the course to be taken and accelerate research and development on SPM fields in future.

I would like to thank all authors for their contributions to this book on "Roadmap of SPM." Editorial works by Mrs. Junko Kobayashi and Dr. Ryuji Nishi are greatly appreciated. I also thank the Japan Society for the Promotion of Science (JSPS) for the support of this "SPM Roadmap" project and Springer-Verlag for their fruitful collaboration. It is hoped that this book will accelerate SPM research and development further, and that "SPM Roadmap" project becomes global as "THE INTERNATIONAL TECHNOLOGY ROADMAP FOR SEMICONDUCTORS (ITRS)."

Osaka, August 2006   *Seizo Morita*

Contents

8 STM-Induced Photon Emission Spectroscopy

9 Scanning Atom Probe

10 Chemical Discrimination of Atoms and Molecules

11 Manipulation of Atoms and Molecules

List of Contributors

Rehana Afrin
Laboratory of Biodynamics
Graduate School of Bioscience and
Biotechnology
Tokyo Institute of Technology
4259 Nagatsuta, Midori-ku
Yokohama 226-8501, Japan
rafrin@bio.titech.ac.jp

Toshio Ando
Department of Physics
Kanazawa University
Kakuma-machi
Kanazawa 920-1192, Japan
tando@kenroku.kanazawa-u.ac.jp

Yasuo Cho
Research Institute of Electrical
Communication
Tohoku University
2-1-1 Katahira, Aoba-ku
Sendai, Miyagi 980-8577, Japan
cho@riec.tohoku.ac.jp

Takashi Furukawa
Advanced Technology Research
Department
Central Research Laboratory
Hitachi, Ltd.
1-280, Higashi-koigakubo
Kokubunji, Tokyo 185-8601, Japan
takashi.furukawa.jh@hitachi.com

Atsushi Ikai
Laboratory of Biodynamics
Graduate School of Bioscience and
Biotechnology
Tokyo Institute of Technology
4259 Nagatsuta, Midori-ku
Yokohama 226-8501, Japan
aikai@bio.titech.ac.jp

Shin-ya Hasegawa
Storage & Intelligent Systems
Laboratories
Fujitsu Laboratories Ltd.
10-1 Morinosato-Wakamiya
Atsugi, Kanagawa 243-0197, Japan
hasegawa.shinya@jp.fujitsu.com

Shuji Hasegawa
Department of Physics
School of Science
University of Tokyo
7-3-1 Hongo, Bunkyo-ku
Tokyo 113-0033, Japan
shuji@surface.phys.s.u-tokyo.
ac.jp

Sumio Hosaka
Department of Nano-Material
Systems (NMS)
Graduate School of Engineering
Gunma University

1-5-1 Tenjin
Kiryu 376-8515, Japan
hosaka@el.gunma-u.ac.jp

Yasushi Kadota
Quality of Management Division
Reliability-Engineering-Office
Ricoh Company, LTD.
810 Shimo-imaizumi,
Ebina, Kanagawa 243-0460, Japan
yasushi.kadota@nts.ricoh.co.jp

Tadahiro Komeda
Institue of Multidisciplinary
Research for Advanced Materials
Tohoku University
Katahira 2-1-1, Aoba-ku
Sendai 980-8577, Japan
komeda@tagen.tohoku.ac.jp

Osamu Kubo
Electro-Nanocharacterization Group
Nanomaterials Laboratories
National Institute for Materials
Science
1-1 Namiki
Tsukuba, Ibaraki 305-0044, Japan
KUBO.Osamu@nims.go.jp

Seizo Morita
Department of Electrical
Electronic and Information
Engineering
Graduate School of Engineering
Osaka University
Yamada-Oka 2-1
Suita 565-0871, Japan
smorita@ele.eng.osaka-u.ac.jp

Tooru Murashita
NTT Photonics Laboratories
NTT Corporation
3-1 Morinosato-Wakamiya
Atsugi 243-0198, Japan
murasita@aecl.ntt.co.jp

Yoshitsugu Nakagawa
Toray Research Center, Inc.
3-3-7 Sonoyama
Otsu 520-8567, Japan
Yoshitsugu_Nakagawa@trc.toray.
co.jp

Ken Nakajima
Tokyo Institute of Technology
2-12-1 O-okayama, Meguro-ku
Tokyo, 152-8552, Japan
knakaji@polymer.titech.ac.jp

Masakazu Nakamura
Department of Electronics and
Mechanical Engineering
Faculty of Engineering
Chiba University
1-33 Yayoi-cho, Inage-ku
Chiba 263-8522, Japan
nakamura@faculty.chiba-u.jp

Osamu Nishikawa
Kanazawa Institute of Technology
7-1 Ohgigaoka Nonoichi
Ishikawa 921-8501, Japan
nisikawa@neptune.kanazawa-it.
ac.jp

Hiroshi Onishi
Department of Chemistry
Faculty of Science
Kobe University
1-1 Rokko-dai-cho, Nada
Kobe 657-8501, Japan
oni@kobe-u.ac.jp

Toshiharu Saiki
Department of Electronics and
Electrical Engineering
Keio University
3-14-1 Hiyoshi, Kohoku
Yokohama 223-8522, Japan
saiki@elec.keio.ac.jp

Yasuhiro Sugawara
Department of Applied Physics
Graduate School of Engineering
Osaka Univesity
2-1 Yamadaoka,
Suita 565-0871, Japan
sugawara@ap.eng.osaka-u.ac.jp

Shukichi Tanaka
Kobe Advanced ICT Research
Center
National Institute of Information
and Communications Technology
588-2 Iwaoka, Nishi-ku
Kobe 651-2492, Japan
tanakas@nict.go.jp

Masahiko Tomitori
School of Materials Science
Japan Advanced Institute of Science
and Technology
1-1 Asahidai
Nomi, Ishikawa 923-1292, Japan
tomitori@jaist.ac.jp

Masaru Tsukada
Graduate School of Science and
Engineering

Waseda University
513 Waseda Tsurumaki-cho, NTRC
bldg 120-5,
Shinjuku-ku, Tokyo 162-0041, Japan
tsukada@cms.nano.waseda.ac.jp

Koji Usuda
Corporate Research & Development
Center
TOSHIBA Corporation
1, Komukai Toshiba-cho, Saiwai-ku
Kawasaki 212-8582, Japan
koji.usuda@toshiba.co.jp

Hirofumi Yamada
Department of Electronic Science
and Engineering
Kyoto University
Nishikyo, Kyoto 615-8510, Japan
h-yamada@kuee.kyoto-u.ac.jp

Masamichi Yoshimura
Graduate School of Engineering
Toyota Technological Institute
2-12-1 Hisakata, Tempaku
Nagoya 468-8511, Japan
yoshi@toyota-ti.ac.jp

Science and Technology in the Twenty-First Century

Seizo Morita

1.1 Trend of Science and Technology in the Twenty-First Century

Since the end of the twentieth century, technology has been developing significantly. Personal mobile communication began with pager, then personal handyphone system (PHS), and now into cellular phone. Cellular phone has already become a third-generation system, and one can photograph, film, e-mail, and even surf the Net. Cellular phones will soon become a wearable ubiquitous computer, which can exchange information anytime and anywhere. Various dream technologies in the electronics field are hoped to be realized within 20 years as listed in Table 1.1. These were predicted by the eighth research on the prospect of technical development by the Ministry of Education, Culture, Sports, Science and Technology of the Japanese government (MEXT) in 2005. Accelerated development of information and communication fields based on rapidly developing electronic devices is the driving forces for realization of such dream technologies.

The backbone of rapid progress and development of electronics field is the accelerated miniaturization of semiconductor integrated circuits. Moore's law proposed by Dr. Moore, cofounder of Intel, predicts that "Integration scale of semiconductor chip will become four times for every three years (two times for every 18 months) approximately," and also "Performance of microprocessor will become four times for every three years (two times for every 18 months) approximately," so that the integration scale of semiconductor chip and performance of microprocessor will develop exponentially. By miniaturization, in the case of Intel, cost of one transistor dropped from more than one dollar in 1965 to one ten-thousandth of one cent at 2005, i.e., one part of million after 40 years. This drop can be compared with that of the TV at ten thousand dollars, which dropped to one cent after 40 years. Besides, by miniaturization, in the case of Intel, the performance of microprocessor improved from about 1.5 MIPS (MIPS: million instructions per second) at 1979 to more than

Table 1.1. Prospected year of technical realization in electronics fields

Prospected year of technical realization	Subject
2011	A system that can predict and judge troubles and accidents, with various kinds of sensors distributed in automobiles
2012	An ubiquitous computer made of one chip that can exchange information anytime and anywhere
2015	An artificial intelligent chip that can understand feelings from an individual's face
2015	One maid robot in every home that can clean up and do laundry
2015	A sensor that can detect diastrophism and predict earthquake several minutes before its occurrence
2018	A microrobot in the human body that can examine inward, with sensors/controllers/actuators integrated by micromachine technology
2024	An information machinery that can calculate 1,000 times faster than C-MOS logic circuit in cases of particular use, based on quantum-computing principle

10,000 MIPS in 2005, i.e., 6,667 times after 26 years. This improvement can be compared with the running distance record in 1 s. was improved from 1 m (the 100-m record is 100 s) in the first year to 6,667 m (the 100-m record is 0.015 s) after 26 years. Half-pitch of line and space of dynamic random-access memory (DRAM) semiconductor memory was 80 nm in 2005. As a result, the number of transistors on one Si chip increased up to several hundred millions. To enable proper functioning of all transistors on one Si chip, fluctuation of fine patterning should be less than 10% of pattern (8 nm in the case of half-pitch pattern). Besides, to precisely evaluate the accuracy of fine patterning, measurement error of fine patterning should be less than 1% of pattern (0.8 nm in case of half-pitch pattern). Thus, rapid miniaturization of large-scale integrated circuit (LSI) requires rapid improvement of spatial resolution of measurement and characterization. As a result, requirement in 2005 was 0.8 nm, i.e., single angstrom (= atom scale) region. Further, improvement of spatial resolution makes the measuring area increasingly smaller, so that the signal level for measurement and characterization becomes correspondingly weaker. Hence, for the improvement of spatial resolution, we need improvement of the measurement sensitivity (signal-to-noise ratio).

The role of scanning probe microscope (SPM), which is the tool for observation and characterization with the spatial resolution of single nanometer and subnanometer (= atomic scale =angstrom), is becoming increasingly important because of the progress of rapid miniaturization. In this book, we introduce our future prospect on SPM research and development, i.e., "SPM roadmap 2005," related to the improvement of better sensitivity, better performance, greater functionality, better spatial resolution, and higher measurement speed, which can measure various physical properties of diverse materials in wider varieties of environments. In Chaps. 2–4, we introduce our future prospect on major SPMs such as scanning tunneling microscope (STM), atomic force microscope (AFM) and near field optical microscope (NSOM). In Chaps. 5–9, we introduce our future prospect on the secondary SPM, i.e., the SPM family except for the major SPM. In Chaps. 10–17, we introduce our future prospect on the emerging and growing techniques related to SPM. In Chaps. 18–22, we introduce our future prospect on the application fields of SPM. In Chap. 23, we introduce our future prospect on the theory and simulation of SPM. Finally, in Chap. 24, we discuss the future prospect on SPM.

1.2 Previous Prospect in SPM Roadmap 2000 and the State-of-the-Art

"SPM Roadmap 2005" is the second future prospect on SPM following the previous "SPM Roadmap 2000," which was published in Japanese with the title of "Scanning Probe Microscope – Its Basis and Future Prospect" on February 10, 2000. We have listed a part of the previous future prospect in SPM in Table 1.2. Hereinafter, we evaluate the previous future prospect "SPM Roadmap 2000" by comparing it with the current state (as of 2005).

In the last five years after the release of previous future prospect "SPM Roadmap 2000," atomic force microscope (AFM) has become one of the most rapidly developed SPMs. As shown in Table 1.2, one of the future prospects

Table 1.2. Future prospect in SPM roadmap 2000 and the result

Microscopes and others	Subject	Prospect in SPM Roadmap 2000	Note (the achieved year, etc.)
AFM	Fundamental experiment of atom manipulation with AFM	2003	2002 (atom extraction by vertical atom manipulation)
AFM	Manipulation of atoms and molecules with AFM	2010	2003 (removal and repair by vertical atom manipulation) 2005 (lateral atom manipulation) 2005 (embedded atom letters [atom inlay])
AFM	True atomic resolution under liquid environment	2010	2005 (mica)
Tip	Sample shipment of carbon nanotube tip	2000	2001

on AFM, "Fundamental experiment of atom manipulation with AFM," was achieved in 2002, 1 year earlier than the future prospect in "SPM Roadmap 2000," where Si adatom in Si(111)7×7 at a low temperature of 9.3 K was mechanically extracted by the vertical atom manipulation using the mechanical vertical contact between the tip apex Si atom and the surface Si adatom [1]. Besides, another future prospect on AFM, "Manipulation of atoms and molecules with AFM," was achieved in 2003 with full reproducibility, seven years earlier than the future prospect in "SPM Roadmap 2000," where Si adatom in Si(111)7×7 at a low temperature of 80 K was mechanically removed and then repaired (deposition of Si atom from the tip apex to the created Si adatom vacancy) by the vertical atom manipulation using the mechanical vertical contact between the tip apex Si atom and the selected surface site [2]. Further, in 2005, a single atom adsorbed on Ge(111)-c(2×8) substrate was laterally manipulated one by one along $[1,\bar{1},0]$ crystal axis using the raster scan method [3]. Moreover, in the same year, lateral atom-interchange manipulation phenomenon, which interchanges the embedded heterogeneous atom such as the selected Sn adatom embedded in Ge(111)-c(2×8) substrate with the selected adjacent Ge adatom, was discovered at room temperature (RT) [4]. Such tip-induced lateral atom-interchange manipulation was understood as the tip-induced directional thermal diffusion of the selected atom. Using such a newly discovered lateral atom manipulation method that can be applicable even to the system consisting of multiatom species, "Atom Inlay" ("that is the embedded atom letters "Sn" (the symbol of tin atom) consisting of 19 Sn atoms embedded in Ge(111)-c(2×8) substrate"), was successfully created as shown in Fig. 1.1 at RT [4]. Besides, one of the future prospects on AFM, "True atomic resolution under liquid environment," was achieved in 2005, five years earlier than that in "SPM Roadmap 2000" [5]. Such rapid progresses of AFM at least partly owe to the stimulation by the future prospect in "SPM Roadmap 2000."

1.3 SPM Roadmap

1.3.1 Roadmap from Both Sides of Seeds and Needs

There are two directions of the SPM roadmap. One is the SPM roadmap from the viewpoint of SPM user (needs' side), while the other is the SPM roadmap from the view point of SPM developer (seeds' side). Chapters 2–17 treat the SPM roadmap on the development of SPM instruments and related techniques from the view point of SPM developer (seeds' side). Chapters 18–22 treat mainly the SPM roadmap from the viewpoint of SPM user (needs' side), but partly from the view point of SPM developer (seeds' side). The SPM roadmap from the view point of SPM developer (seeds' side) discusses the subject, that is, "Until when will such an affair become possible technically?" The SPM roadmap from the viewpoint of SPM user (needs' side) discusses the subject,

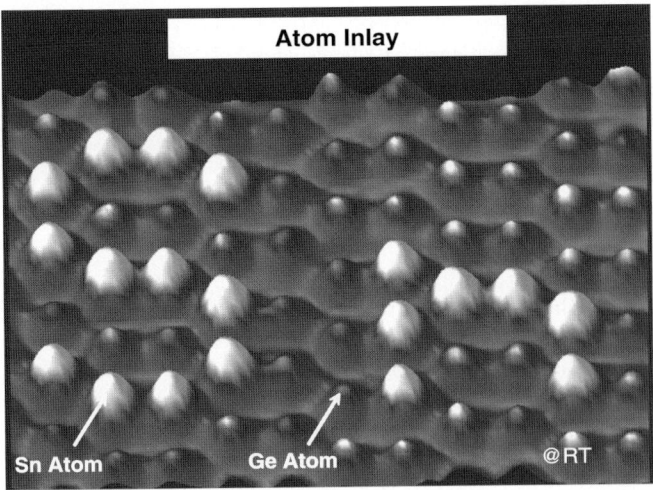

Fig. 1.1. The first "Atom Inlay," that is, atom letters figured by 19s of Sn atoms embedded in Ge(111)-c(2×8) at room temperature

that is, "Until when should such an affair become possible technically?". As a result, there will be some difference in future prospects depending on the view points of either SPM developer (seeds' side) or SPM user (needs' side). Hence, the reader has to be careful of which side the SPM roadmap is written.

1.3.2 Various Directions of SPM Roadmap

There are various directions in SPM roadmap. Pure SPM developer (pure seeds' side) will aim at "Better constituent technologies and instruments" such as better performance, better sensitivity (e.g., improvement of signal-to-noise ratio), higher spatial resolution, more precise control (e.g., tip–surface distance), higher measurement speed, greater functionality, and combined SPM with more functions, while applied SPM developer (applied seeds' side) will aim at usage in wider environments (e.g., in gases and liquids), chemical identification of atoms and molecules and atom manipulation/assembly. SPM roadmap of applied seeds' side includes the viewpoints from not only seeds' side but also needs' side. As a result of such variations in SPM roadmap direction, the SPM roadmap may become a polygon. For example, in the case of the seeds' side, Chap. 9 scanning atom probe (SAP) has six axes namely, detection efficiency, mass resolution, spatial resolution, multiplicity, sample for analysis, and theoretical analysis. In the case of needs' (user) side, Chap. 19 SPMs characterization of LSI devices also has six axes such as microscopic shape, crystal dislocation, contamination by dopant impurities, electric characteristics, chemical states, and stress. Roadmap of all SPMs such as AFM is intrinsically multiaxis. Important axis and/or alternative of AFM, however, are introduced in different chapters such as Sect. 10.2 "Chemical Identification of Atoms and molecules by AFM," Sect. 11.2 "Manipulation of Atoms and Molecules by AFM," Chap. 12 "Multi-Probe SPM," Chap. 13 "AFM Measurement in Liquid" and Chap. 14 "High Speed AFM".

References

1. S. Morita, R. Wiesendanger, E. Meyer (eds.), *Noncontact Atomic Force Microscopy* (Springer, Berlin Heidelberg New York 2002) 3.10
2. N. Oyabu et al., Phys. Rev. Lett. **90**, 176102 (2003)
3. N. Oyabu et al., Nanotechnology **16**, S112 (2005)
4. Y. Sugimoto et al., Nature Materials **4**, 156 (2005)
5. T. Fukuma et al., Rev. Sci. Instrum. **76**, 053704 (2005)

2

Scanning Tunneling Microscopy

Masahiko Tomitori

2.1 Basic Principle of Scanning Tunneling Microscopy

When an atomically sharpened tip is brought closer to a sample surface, e.g., $\sim 1\,\mathrm{nm}$, under a bias voltage between the tip and a sample, a small electric current starts to pass between them before they are in contact. This is the so-called quantum tunneling effect; we refer to the current as tunneling current. The tunneling current drastically increases with decreasing separation between the tip and the sample. Conversely, we can precisely evaluate the change in the separation by measuring the tunneling current. Thus, if we scan the tip over the sample surface while keeping the tunneling current constant, the tip movement depicts the surface topography, because the separation between the tip apex and the sample surface is always constant. This is the basic idea to develop a novel surface imaging microscope with atomic resolution, that is, scanning tunneling microscopy (STM) [1]. Although the STM is really a simple instrument basically consisting of a tip and a sample that are faced each other as shown in Fig. 2.1, the obtained resolution is so high as the individual atoms can be resolved when the tip apex is atomically sharp.

Here, we look at the change in tunneling current with the separation on the basis of a one-dimensional tunneling barrier that is sandwiched as a metal–vacuum–metal junction. We assume that the electronic density-of-state (DOS) at the tip apex is constant with respect to the energy of electron, and consider only the electron tunneling process without any energy loss, i.e., the elastic tunneling process. By approximately calculating the tunneling probability, the tunneling current I can be expressed as

$$I \propto \exp\left(-2s\sqrt{\frac{2m}{\hbar^2}\left(\langle\phi\rangle - \frac{e\,|V|}{2}\right)}\right). \tag{2.1}$$

Here, s is the separation between the tip apex and the sample surface, m the electron mass, e the electron charge, $\hbar$ the Plank constant, $\langle\phi\rangle$ the averaged work function of the tip and the sample, and V the bias voltage. For example,

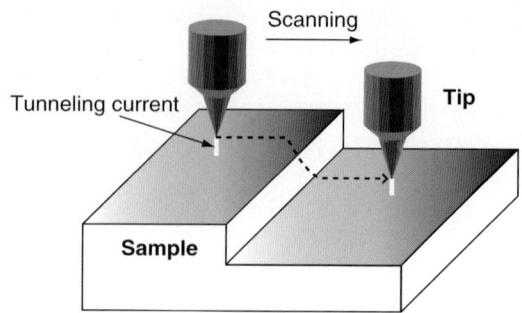

Fig. 2.1. Principle of a scanning tunneling microscope

when $\langle\phi\rangle - e\,|V|\,/2$ is 5 eV, s is 1 nm and changes by 0.1 nm, then I changes by one order of the magnitude. This is the reason why the STM can achieve atomic resolution in the vertical direction z with respect to the sample surface; a vertical resolution of better than 0.01 nm can be obtained. In addition, a lateral resolution of about 0.2 nm is also achievable, which crucially depends on the sharpness of the tip.

Here, it is worth noting that what we do in the STM is to convert the spatial change in tunneling current into topographic image. Precisely speaking, the tunneling current changes with the surface electronic states of the tip and the sample as well as the separation between them. The states spread from their surface to vacuum within peculiar regions, which change with their electronic energies. Therefore, the tunneling current changes with applied bias voltage, the electronic wavefunctions at the apex, i.e., the atomic species at the tip apex, and so on. This indicates that we always have to take care of the interpretation of STM images. On the contrary, these complexities of the STM can be used beneficially to measure the electronic properties of the sample with high spatial resolution, leading us to spectroscopies based on the STM.

From an instrumental point of view, the STM is chiefly composed of the following parts:

(1) An atomically sharpened tip, and a sample that is placed to be faced to the tip apex.
(2) Positioning systems to select the target area on a sample surface to be observed, control the separation between a tip and a sample, and scan the tip over the area, coarsely in a range of $\sim$ mm, and finely with atomic resolution.
(3) A current amplifier to measure the tunneling current and a feedback circuit to keep the tunneling current constant.
(4) A computer system to generate an x–y scanning signal, to record the tip trace over a sample surface and output an STM image through computer graphic processes.

(5) A vibration isolation system to prevent the disturbance from being transmitted from lab environments, not to change the separation mechanically. An air- or metal-spring suspension bench with dampers, metal-rubber stacks, and so on is of use. In addition, an STM head is sometimes placed in a sound isolation box.

(6) An environmental control system. For example, an ultrahigh vacuum (UHV) chamber and pumps to keep the tip and the sample clean, a clean gas purging system, a liquid cell with an electrochemical control, temperature controls for high- and low-temperature observations, including a cryostat, and so on.

Figure 2.2 shows a block diagram of an STM system. A tip is attached at a corner of a scanner, consisting of three rectangular rods of piezo ceramics (Pb(Zr,Ti)O$_3$ (PZT)) that are crossing perpendicularly. The PZT rod can be elongated with increasing voltage applied between two electrodes on its opposite longitudinal faces. For example, the rod elongates 1–2 nm per 1 V. To scan the tip faster, a scanner with a higher mechanical resonance frequency is demanded; a compact and tube type of piezo scanner, a shear piezo scanner, and so on were developed.

A tunneling current less than the order of nA in magnitude is detected by a current amplifier with a conversion ratio of 10^{7-9} V A^{-1}. The output of the current amplifier is fed into an absolute-logarithmic amplifier to linearize the relation between the tunneling current and the separation between the tip and the sample. Afterward, a reference value I_{ref}, precisely log I_{ref}, is subtracted from the linearized signal, which is a target value for the STM feedback operation to keep the current constant. Then, the signal is input to the feedback control, in which we can select a suitable set of gain and time constant to keep the current constant in a stable manner. Finally, the output from the feedback control is amplified with a high-voltage amplifier

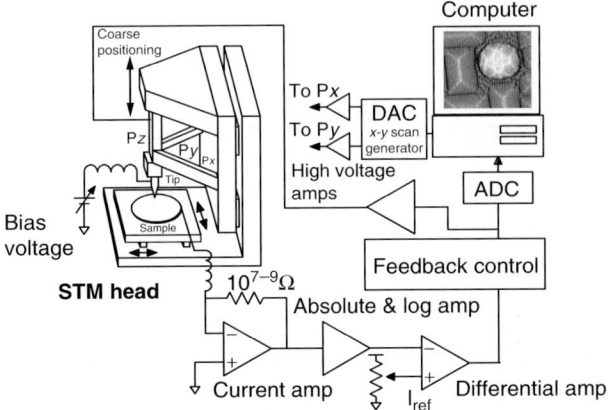

Fig. 2.2. Block diagram of STM

with an output range higher than ~100 V, which is applied to the z piezo. Here the loop is closed for the feedback operation; when the tunneling current exceeds the target value, the feedback control retracts the tip, and conversely, when the tunneling current decreases than that, the control brings the tip closer to the sample. To observe an STM image, x–y piezos are scanned by changing voltages applied to them in saw-like waveforms that are generated by a computer with digital-to-analog converters (DACs). The signal output from the feedback control is fed into an analog-to-digital converter (ADC) installed in the computer. The STM image, processed from the three-dimensional data of x–y–z voltages applied to the scanner, is displayed on a computer monitor.

When we start the STM scanning, to obtain the tunneling current, we should bring the tip closer to the sample from a farther separation as a few mm using a coarse positioning system. Several types of coarse positioning systems have been developed:

(1) The so-called *inchworm* or *louse*, consisting of piezo ceramics as a main drive and clamping feet to fix the body on a base plate [1].
(2) Fine screw mechanism, sometimes combined with a reduction utilizing a lever mechanism [2, 3].
(3) Inertia motion stage utilizing a mechanical impact induced by quick motion of a piezo with a weight. This can proceed when the impact exceeds the friction between its body and a base plate [4, 5].

2.2 History of STM

The STM was invented by Binnig, Rohrer, and their co-workers at IBM Research Laboratory, Zurich, Switzerland, in the early 1980s. They had tried to measure the spatial distribution of tunneling current between a tip and a superconducting sample at a separation of about 1 nm by scanning the tip, which was inspired by extending the point contact tunneling spectroscopy for characterizing superconducting properties. During their aggressive study for this target, they significantly recognized the possibility of a microscope with tremendous high resolution on an extended line of this development of the scanning tip (probe) method. Their scientific achievement of an atomic resolution using the STM based on their original idea should be greatly admired, and triggered the opening of nanotechnology, which has spread explosively all over the world since the end of the twentieth century. The Royal Swedish Academy of Sciences had awarded the 1986 Nobel Prize in Physics to Binnig and Rohrer for their design of the scanning tunneling microscope.

It is noted that a microscope having a scanning probe mechanism, almost the same as the STM named *topographiner*, had been developed by Young et al. in 1972 [6]. They utilized a sharpened metal tip operated in a field emission mode, where a high negative voltage is applied to the tip to extract electrons from the apex, and scanned the tip over a sample surface with a feedback system combined with a piezo scanner while keeping the field emission current constant. Although they succeeded in obtaining surface topography on a nanometer scale and tried to detect tunneling current by bringing the tip close to the sample as ~1 nm, they had not achieved the atomic resolution using the topographiner. In those days, fundamental equipment to achieve the nanometer scale mechanical measurement seemed to be insufficient. On the other hand, Binnig and Rohrer had successfully developed a stable vibration isolation stage vigorously, and distinctively aimed at the operation in a tunneling regime from the beginning of the development. However, the development of topographiner by Young should be highly evaluated as a pioneer in this field; it is noted that the Royal Academy commented on Young's contribution in the award press release. By the mid-1980s the STM development had been spread to famous laboratories, mainly in Europe and the USA, and the high performance of STM had been confirmed. In 1986, Binnig et al. also successfully developed an atomic force microscope (AFM) based on the STM (see in detail in Chap. 3). While the motivation of AFM development was initiated to measure the force between an STM tip and a sample, it is noticeable that the AFM can be applied to insulating materials. This feature has not been realized by the other high-resolution microscopes. Recently, the application area of AFM has been much wider than that of the STM.

Almost all trials to extend the performance of STM had been demonstrated by the late 1980s. For example, the measurement of the spatial change in apparent work function was done by modulating the separation between the tip and the sample and by detecting the modulated tunneling current using a lock-in amplifier. It is noted that the apparent work function is different from the work function in traditional surface science; it means the spatial correlation of overlapping between the wavefunctions of the tip and the sample. There is a method to depict the spatial change in surface electronic density by modulating the bias voltage while scanning the tip. Scanning tunneling spectroscopy (STS) was also developed to deduce the distribution of surface electronic density of states with respect to the electronic energy with atomic resolution by recording current versus bias voltage (I–V) curves while holding the tip–sample separation intermittently during scanning [7]; which is also referred to as *current imaging tunneling spectroscopy* (CITS). Furthermore, by taking the second derivatives of I–V curves, inelastic tunneling processes were evaluated to survey vibrational characteristics of molecules adsorbed on sample surfaces [8] (see Sect. 10.1). In addition, the excitation of electron standing waves in a vacuum gap between a tip and a sample [9], the spectroscopic detection of light emission and secondary electron emission from tip–sample regions [10, 11] had been demonstrated. Surface modification and atomic manipulation using the STM had started since the mid-1980s as well (see Chap. 11). In the 1990s commercial available STMs had emerged, including STMs operated at low temperatures and in UHV, that have been widely used all over the world. The STMs combined with a scanning electron microscope (SEM) or with a transmission electron microscope (TEM) have also been used widely.

2.3 Present States and Unsettled Issues

At present the application of STM has exceedingly been expanded as a tremendous high-resolution surface microscope in materials science for samples with electric conductivity, and the STM will continue to be used in various fields. On the other hand, the problem to prepare an atomically sharpened well-defined tip has still remained. As mentioned above, the spatial resolution of STM is governed by the sharpness of the tip apex; ideally a tip having one atom at the apex with a stable atomic arrangement is required. Furthermore, since the tunneling current changes with the electron wavefunctions of the tip, it is indispensable to control atomic species at the tip apex, including the atomic structure supporting the apex atom. Of course, the structure should be stable during the scanning. Practically, until an observed image obtains a preferable resolution, one sometimes tries to bring a tip in contact to a sample, or applies a high voltage to the tip to induce a change in the atomic structure at the apex owing to very high electric field. When the STM is placed in UHV, since a field emission microscope (FEM) and a field ion microscope (FIM) can image the tip apex [12], evaluation of tips with the FEM/FIM is very helpful to prepare a high-quality tip in the same chamber with an STM. In situ tip observation and modification using an STM combined with a TEM or an SEM are also of use. Furthermore, an in situ tip growth with application of high voltage while the tip is being faced to the sample [13], thermal field treatments [14], a nanopillar growth on a tip by pulling it from a heated sample with an SPM [15], and so on have been proposed. In addition, attaching a carbon nanotube to the tip apex has attracted much interest. The tip should be heated before use to remove contamination layers covering the apex, which may cause unstable imaging operation. It is suggested that preparation of the tip should be designed depending on the experimental purpose; consider which properties of a sample would be enhanced by STM observation with the prepared tip. It is also noted that a shape around a tip end except its apex could limit the image quality for rough sample surfaces; a rugged tip end generates many artifacts such as ghosts with respect to true topography in observed STM images.

As for other issues, attention to nonlinear expansion of piezo ceramics with creeping and hysteresis with respect to applied voltage for scanning should be paid. In order to measure the surface feature with a high precision, displacement sensors to calibrate the expansion of piezos should be installed. Moreover, the feedback controls with the sensors are also implemented to linearize the expansion in real time. In addition, to observe a small region for a long time, the mechanical thermal drift of an STM head should be reduced, and the mechanical response time for fast scanning should be improved.

2.4 Roadmap

As a roadmap of STM, the following issues are listed (see Fig. 2.3).

2.4.1 Further STM Combination with Other Microscopies and Spectroscopies

The STM combination with other microscopies and spectroscopies is really useful, which will be greatly promoted together with the development of the STM/STS related techniques. The high spatial resolution of STM can fascinatingly cover up the faults of other powerful analytical instrumentations. In fact the combination with a TEM and an SEM provides us with complementary information on the tip positioning and the sample surfaces, which seem to have advantages on measurements of current–voltage characteristics, atom manipulation, and so on. For example, the TEM/SEM combination with STM/STS/nc-AFM (see Chap. 3) sounds extremely attractive, where simultaneous measurements of force and tunneling or pseudo-contact current will be performed with atomic resolution while observing the tip and the sample with the TEM/SEM. In addition, the combination with surface spectroscopic techniques operated in UHV seems easy technologically, in which the tip can work as an electron source for electronic state excitation or photon emission; some of them have already been demonstrated.

2.4.2 Miniaturized, Multi-, and Intelligent STM

In general, by miniaturizing an STM head, the mechanical resonance frequency of the head increases, leading to realization of fast scanning with increasing tolerance against the vibration transmission from the surrounding environment. Miniaturization of STM using semiconductor lithographic techniques has already been reported. With the miniaturization, the multi-tip

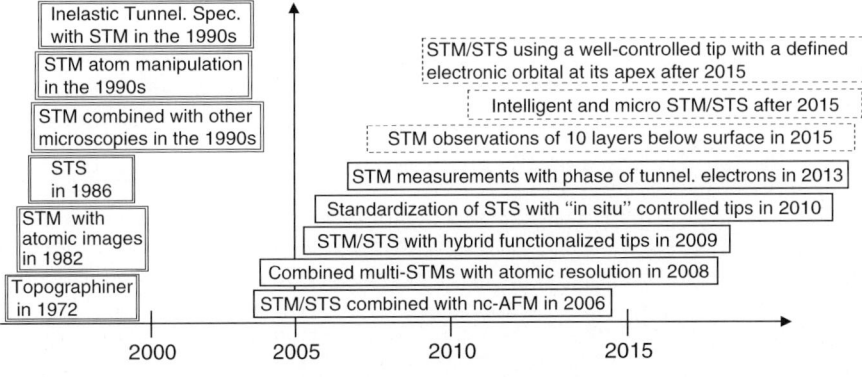

Fig. 2.3. Roadmap of STM

capability of STM can be demonstrated easily. In addition, if the intensive miniaturization, including STM control circuits is achieved, the intelligent STM that can communicate with an observer and works according to the observer's order with a capability of self-judgment will not be a dream.

2.4.3 Well-Defined Tips and Hybrid Tips

As mentioned above, the preparation of a well-defined atomically sharpened tip is one of the most demanding issues for any SPMs. In future it is indispensable to grow the tip in situ in an atomically controlled manner and prepare functionalized tips with specific physical and chemical capabilities as demanded. The technique to control the electron orbitals at the tip apex is a final goal.

2.4.4 Evolution of STM Utilizing Phase of Tunneling Electron and Ballistic Electron

The STM utilizes the absolute value of tunneling current. On one hand, the coherence of tunneling electrons is high, since they have a narrow energy distribution. Using this high coherence of tunneling electrons, a novel analytical method may be developed in future; for example, to exploit the inner structure of a sample up to 10 layers below the surface, including atomic arrangements and species, based on electron scattering on an atomic scale. Ballistic electrons from a tip, which are emitted with increasing bias voltage over the work function, is also hopeful to investigate them. The other STM possibilities of atom manipulation, discriminating atomic species and measuring spins are referred to Chap. 10.

References

1. G. Binnig et al., Phys. Rev. Lett. **49**, 57 (1982)
2. P.K. Hansma et al., Science **242**, 209 (1988)
3. J.E. Demuth et al., J. Vac. Sci. Technol. A **4**, 1320 (1986)
4. D.W. Pohl, Rev. Sci. Instrum. **58**, 54 (1987)
5. R.M. Möller, A. Esslinger, M. Rauscher, J. Vac. Sci. Technol. A **8**, 434 (1990)
6. R. Young, J. Ward, F. Scire, Rev. Sci. Instrum. **43**, 999 (1972)
7. R.J. Hamers, R.M. Tromp, J.E. Demuth, Surf. Sci. **181**, 346 (1987)
8. D.P.E. Smith, M.D. Kirk, C.F. Quate, J. Chem. Phys. **86**, 6034 (1987)
9. R.S. Becker, J.A. Golovchenko, B.S. Swartzentruber, Phys. Rev. Lett. **55**, 987 (1985)
10. J.H. Coombs et al., J. Microsc. **152**, 325 (1988)
11. B. Reihl, J.K. Gimzewski, Surf. Sci. **189/190**, 36 (1989)
12. T.T. Tsong, *Atom-probe field ion microscopy* (Cambridge University Press, 1990)
13. S. Heike, T. Hashizume, Y. Wada, J. Vac. Sci. Technol. B **14**, 1522 (1986)
14. M. Tomitori et al., Surf. Sci. **355**, 21 (1996)
15. T. Arai, M. Tomitori, Appl. Phys. Lett., **86**, 073110 (2005)

3

Atomic Force Microscopy

Yasuhiro Sugawara

3.1 Principle

Atomic force microscopy (AFM) [1] is a novel technique for high resolution imaging of conducting as well as nonconducting surfaces. As shown in Fig. 3.1, the physical property sensed in AFM is the interaction force between the sample surface and a sharp probing tip. The cantilevers detecting the atomic force are essential elements to significantly affect the spatial resolution of the AFM. For high resolution imaging of the microscopic structure on the sample surface, very sharp tip on the cantilever is required. For fast imaging, high mechanical resonant frequency of the cantilevers is also necessary. At present, micro-fabricated Si and/or Si_3N_4 cantilevers are widely used. Displacement sensor measures the deflection of the cantilevers with sensitivity better than $0.1\,nm$. As a displacement sensor, optical beam deflection method is widely used, in which angle change of the optical laser beam reflected from the backside of the cantilever is detected.

There are three types of imaging modes of the sample surface in AFM: contact, tapping, and noncontact modes. In the contact mode , the probing tip is always in contact with the sample surface, and surface structure is obtained from the deflection of the cantilever. In the tapping mode, the probing tip is periodically in contact with the sample surface, and surface structure is obtained from the change of the vibration amplitude or phase of the oscillating cantilever [2]. In the noncontact mode, the probing tip is not in contact with the sample surface, and surface structure is obtained from the change of the vibration amplitude or resonant frequency of the oscillating cantilever [3, 4].

In the contact mode, there is a high possibility that the strong repulsive force acting between the sample surface and the probing tip will destroy the sample surface and/or the tip apex. So, at present, the tapping and noncontact modes are widely used because these modes are more gentle than the contact mode.

In the tapping mode, the cantilever is driven at a fixed frequency near resonance with large vibration amplitude. When the probing tip is far from the

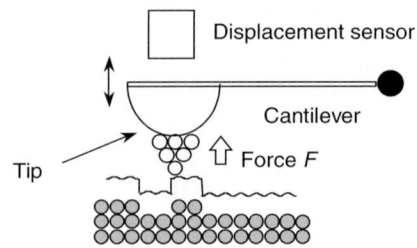

Fig. 3.1. Schematic of AFM

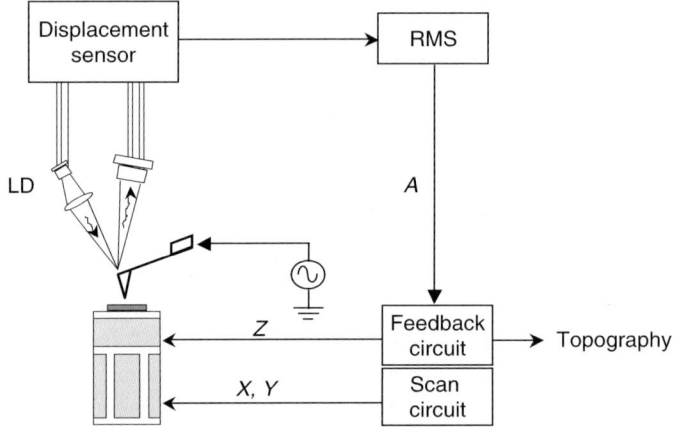

Fig. 3.2. Schematic of tapping AFM

surface, the vibration amplitude of the oscillating cantilever is held constant. When the probing tip is close to the surface, the probing tip is periodically in contact with the sample surface, and the vibration amplitude of the oscillating cantilever decreases due to cyclic repulsive contact between tip and surface with loss of the energy stored in the oscillating cantilever. The surface structure is obtained by maintaining the vibration amplitude at the constant level using the feedback circuit as shown in Fig. 3.2.

We will explain that the loading force acting between the probing tip and the sample surface can be greatly reduced in the tapping mode. We assume that the vibration amplitude of the oscillating cantilever decreases from A_0 to A due to the cyclic repulsive force interaction between the probing tip and the sample surface. Here, if the sample surface does not exit, as shown in Fig. 3.3, the vibration amplitude will increase gradually from A to A_0 with increment per one cycle of [3]

$$\Delta A = \frac{A_0^2 - A^2}{2AQ}, \tag{3.1}$$

where Q is the quality factor of the cantilever. However, actually, the sample surface exists. So, the energy corresponding to the increment of vibration

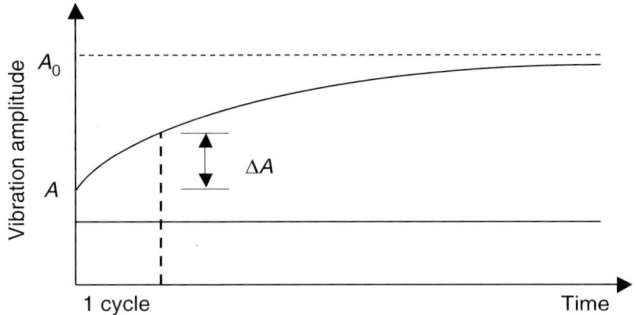

Fig. 3.3. Transition response of vibration amplitude

amplitude ΔA is dissipated by the cyclic repulsive contact between the probing tip and the sample surface. As a result, the vibration amplitude of the oscillating cantilever is kept at the constant amplitude A. In this case, the force acting between the probing tip and the sample surface is roughly given by

$$F = k\Delta A, \tag{3.2}$$

where k is the spring constant of the cantilever. For example, assuming that $A_0 = 20\,\text{nm}$, $A = 10\,\text{nm}$, $k = 1\,\text{N}\,\text{m}^{-1}$, and $Q = 100$, the increment of vibration amplitude ΔA is estimated to be $0.15\,\text{nm}$. This value corresponds to the small loading force of $F = 0.15\,\text{nN}$.

Next, we will explain the basic imaging principle of the noncontact mode AFM. In Fig. 3.4, solid line shows the resonance curve of the cantilever when the probing tip is far from the sample surface without the force interaction. ν_0 is the mechanical resonance frequency of the cantilever. When the probing tip approaches the sample surface, effective spring constant of the cantilever changes from k to $k + \partial F/\partial z$ due to the force gradient $k + \partial F/\partial z$ acting between the probing tip and the sample surface. As a result, the mechanical resonant frequency changes from ν_0 to ν_1 (dotted line in Fig. 3.4). By measuring the frequency shift $\Delta\nu = \nu_0 - \nu_1$, we can estimate the force gradient $k + \partial F/\partial z$.

There are two types of detection methods to measure the frequency shift $\Delta\nu$. One is slope detection method [4] and another is frequency modulation (FM) detection method [5]. In the slope detection method, the cantilever is driven at the off-resonant frequency ν_2. With the force interaction, the vibration amplitude (or phase) of the cantilever at the frequency ν_2 changes. This change ΔA of the vibration amplitude is detected by lock-in amplifier. In AFM imaging, the distance between the probing tip and the sample surface is controlled to keep constant amplitude change ΔA. In the FM detection method, on the other hand, the cantilever is driven at the resonant frequency by the positive feedback system with variable gain amplifier; automatic gain control (AGC) circuit and phase shifter at constant level as indicated in Fig. 3.5. The phase shifter is used to adjust the phase of the feedback system. The frequency

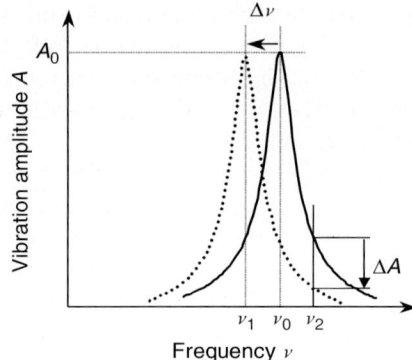

Fig. 3.4. Measurement principle of noncontact AFM

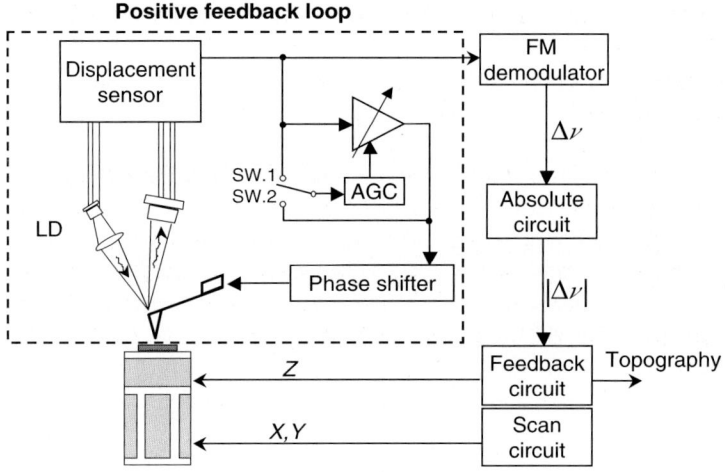

Fig. 3.5. Schematic of FM AFM

shift $\Delta\nu$ of the cantilever is detected by the FM demodulator. In AFM imaging, the distance between the probing tip and the sample surface is controlled to keep constant frequency shift $\Delta\nu$.

3.2 History

In 1986, AFM has been developed into a novel technique for obtaining high resolution images of both conductors and insulators. With the contact mode, for several layered and nonlayered materials, atomic resolution was achieved. However, most of reported data showed either perfectly ordered periodic atomic structure or defects on large lateral scale, but no atomic-scale point defects which were routinely observed by the scanning tunneling microscope

(STM). Under the restricted measurement condition, it was demonstrated that monoatomic step lines and atomic-scale point defects could be observed with atomic resolution. However, the condition for achieving true atomic-scale lateral resolution is extremely restricted due to thermal drifts and signal-to-noise ratio of the force measurement.

In the noncontact mode, on the other hand, the destruction of the initial sharp tip and the sample surface can be avoided. However, for a long time, atomic-scale lateral resolution has not been achieved because of weak distance dependence of the attractive force between the tip and the sample.

In 1993, the tapping mode AFM was proposed, in which the probing tip is periodically in contact with the sample surface, and the surface structure is obtained by maintaining the vibration amplitude at the constant level. In this mode, the tip–sample interaction corresponding to the decrease of the vibration amplitude is mainly the repulsive force interaction with strong distance dependence. Furthermore, the loading force acting between the probing tip and the sample surface can be greatly reduced. As a result, the spatial resolution greatly improved. However, atomic-scale lateral resolution has not been achieved because of the still strong loading force between the probing tip and the sample surface.

In 1995, the noncontact AFM using the FM detection method achieved true atomic resolution [6, 7]. At present, various surfaces such as clean semiconductors, ionic crystals, metal oxide, metal deposited semiconductor, pure metals, and layered material have been observed successively with atomic resolution. Now, the noncontact AFM is expected to become a powerful scientific tool for resolving the atomic features in various fields such as, materials and biological sciences [8].

3.3 Present Situation and Issues

Here, we discuss the various issues of the AFM as an atomic resolution microscopy. The first key issue is the preparation method of the well-defined sharp probing tip, which determines the spatial resolution of AFM images. This issue is the common problem in scanning probe microscopy (SPM). The spatial resolution in the AFM strongly depends on the size of the tip apex. The probing tip detects not only short-range interaction such as chemical force but also long-range interactions such as van der Waals force and electrostatic force. So, very sharp tip apex with a single atom protrusion is required to achieve extreme high spatial resolution. Moreover, it is reported that the pattern of AFM images drastically changes depending on the atom species on the tip apex. So, we have to control an atomic species at the tip apex to identify or recognize atom species on a sample surface using the AFM. In the AFM operating in UHV condition, so far, Ar-ions sputtering method has been widely used to clean the tip apex. However, it is impossible to control an atomic species at the tip apex using the sputtering method. Establishment of the preparation method of the well-defined sharp probing tip is required.

Next key issue is the force sensitivity of the AFM, which also determines the spatial resolution of AFM images. In commercial AFM instrumentations, the optical beam deflection method is widely used as a displacement sensor of the cantilever. However, its force sensitivity is not sufficient to realize the stable atomic resolution imaging and to quantify the very weak force interaction between the tip and the sample surface. Further study to reduce the noise of the displacement sensor and to enhance the force sensitivity is necessary.

3.4 Roadmap

Here, we will predict future new developments of the AFM.

3.4.1 Development of New Force Spectroscopy

It is well known that the force interaction between the probing tip and the sample surface contains atomic-scale on various physical information. For example, mechanical response of the surface structure, charge transfer, energy dissipation, and so on. Such physical information on atoms, molecules, and electrons are very useful to investigate the various phenomena on surfaces. In future, by developing the new force spectroscopy methods using the various stimuli such as microwave, optical beam, X-ray, and magnetic field, such physical information will be investigated with high sensitivity.

3.4.2 Development of AFM Imaging Operating in Special Environments

The AFM utilizes the resonance enhancement of the force sensitivity by vibrating the cantilever at or near the resonance frequency. In a vacuum environment, mechanical Q-factor of the cantilevers is very high and very high force sensitivity is realized. As a result, many successful imaging with atomic and molecular resolution is reported.

On the other hand, in a liquid environment, mechanical Q-factor of the cantilever is greatly reduced due to the viscosity of the liquid, and the force sensitivity is not enhanced very much. As a result, high resolution imaging is very difficult. Various attempts to realize the high force sensitivity in a liquid environment have been performed and true atomic resolution imaging is achieved. In future, by enhancing the force sensitivity further, atomic resolution imaging of hydrophilic and hydrophobic interactions and surface charges on surface are expected in a liquid environment.

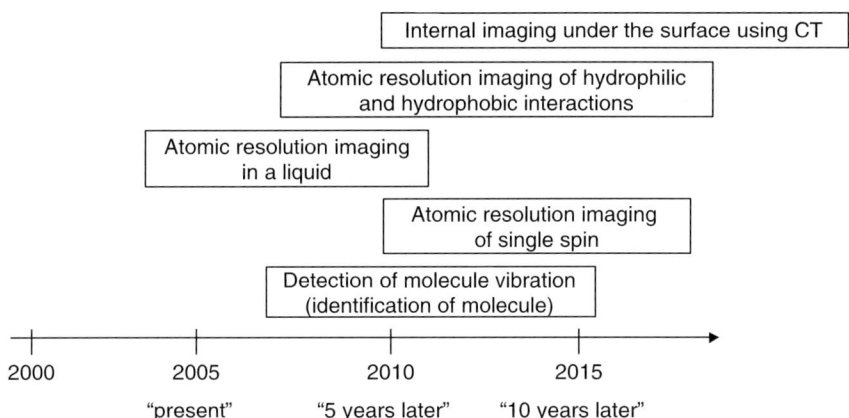

Fig. 3.6. Future predictions for novel AFM techniques

3.4.3 Development of Imaging Method Under the Surface

Usually, in AFM, the surface structure is imaged by using the shot-range force interaction. If we can detect the long-range force from the atoms under the surface separately, it will be possible to visualize the internal structure under the surface and of complex molecules with three-dimensional spatial resolution. Computer tomography (CT) using AFM is hopeful as an imaging method under the surface. We hope that novel method to visualize the internal structure will be developed.

Finally, in Fig. 3.6, we show future predictions for novel AFM techniques. For identification of the atom species, see Chap. 10.

References

1. G. Binnig, C.F. Quate, and Ch. Gerber, Phys. Rev. Lett. **56**, 930 (1986)
2. P.K. Hansma et al., Appl. Phys. Lett. **64**(13), 1738 (1994)
3. Q. Zhong et al., Surf. Sci. Lett. **290**, L688 (1993)
4. Y. Marti, C.C. Williams, and H.K. Wickramasinghe, J. Appl. Phys. **61**(10), 4723 (1987)
5. T.R. Albrecht et al., J. Appl. Phys. **69**(2), 668 (1991)
6. F.J. Giessibl, Science **267**, 68 (1995)
7. Y. Sugawara et al., Science **270**, 1646 (1995)
8. S. Morita et al., Scanning Probe Microscopy; Characterization, Nanofabrication and Device Application of Functional Materials (Kluwer, Dordrecht, The Netherlands 2005)

4

Near-Field Scanning Optical Microscope

Toshiharu Saiki

4.1 Principle of NSOM

The principle of near-field scanning optical microscopy (NSOM) can be simply modeled by the electromagnetic interaction of two very closely positioned nano-objects, which represent a probe and a sample [1]. When the two nano-objects are illuminated with light, the electric dipole moments are induced into both the objects and they couple through the dipole–dipole interaction. Owing to the short-range nature of dipole–dipole interaction, the total dipole moment rapidly changes with relative position of the two objects. By detecting a radiation from the total dipole in the far field with scanning one of two objects, high spatial resolution imaging can be obtained. Roughly speaking, the resolution is determined by the size of object.

In most of measurements with NSOM, inelastic scattering signals like fluorescence or Raman scattering, are more informative. In such cases, it is more intuitive to treat the probe (one of two objects) as a localized light source to illuminate the sample (the other one).

There are two types of NSOM in terms of probe structure as a localized light source: aperture-type and scattering-type NSOM. The physical processes involved in both types of NSOM are identical as described above and the differences are restricted to technological considerations. In the aperture-type NSOM, a small opening on a metal film is utilized to generate a nanoscale light spot. Even when the diameter of the aperture is smaller than the wavelength of light, we obtain a light spot same as the size of the aperture. Because the light diffracts immediately after being ejected through the aperture, the probe apex must be kept in close proximity to the sample surface while scanning (aperture–sample distance is regulated at less than 10 nm). In the actual operation of aperture NSOM, an optical fiber based probe is employed as shown in Fig. 4.1a.

Scattering type NSOM uses a sharpened homogeneous metal tip as a probe (Fig. 4.1b). When the tip is illuminated with the polarization parallel to the tip axis, the electric field is enhanced owing to the large surface charge

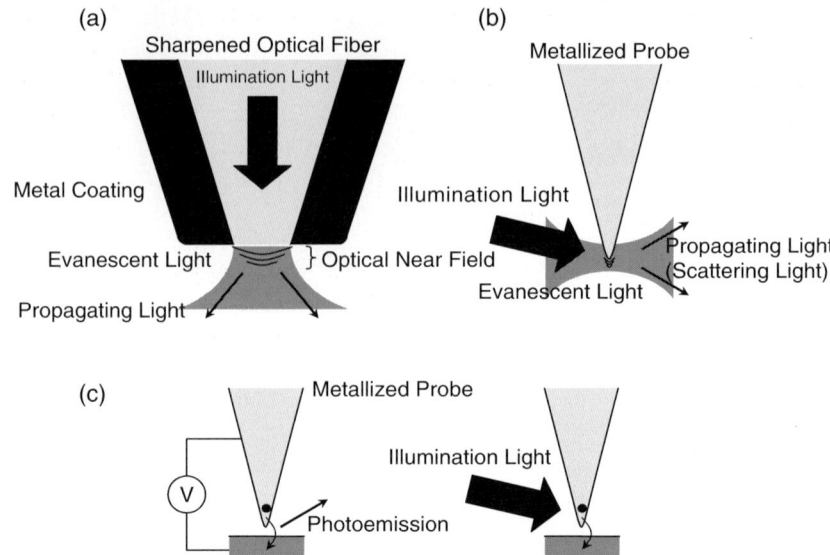

Fig. 4.1. Schematic illustrations of (**a**) aperture NSOM, (**b**) scattering NSOM, and (**c**) photon-emission STM and light-illuminating STM

accumulation at the apex. The spatial resolution defined by the apex diameter is expected. The field enhancement effect can also contribute to the detection of extremely weak signal, such as Raman scattering from single nano-objects. However, because the illumination area is limited by the diffraction to generate unwanted background, elimination of signal from the strong background is critical for imaging with sufficient contrast.

Photon-emission scanning tunneling microscopy (STM) and light-illumination STM, illustrated in Fig. 4.1c, can be categorized into NSOM from a viewpoint of essential contributions of light to them. They allow the atomic resolution originated from STM operation. In photon-emission STM, we detect photons created by inelastic scattering of electrons in the tunneling gap applying a bias voltage. Photon-emission spectra provide electronic state information of nanoscale region, such as a local density of states. Light-illuminating STM measures the tunneling current of photoexcited carriers. Precise spectroscopy of local electronic states can be carried out by means of highly controlled light. By using polarized photons one can couple to a specific spin state. Ultrafast optical pulses enables to observe electron dynamics with a high spatiotemporal resolution.

4.2 Progress in Fundamental Performance of NSOM

The exploration of visible NSOM based on the aperture-type probe started at the beginning of 1980s. Through basic researches by several groups, NSOM began to attract wide and general interests in the early 1990s. Applications to optical data storage and spectroscopy of single molecules and semiconductor quantum structures stimulated the NSOM research field. These works triggered various further applications and technical improvements of NSOM.

Regarding spatial resolution achieved in the 1990s, experiments demonstrating a few nm resolution were reported. However, many of these results suffered from topographic artifacts, i.e., the high resolution of force microscopy for tip–sample distance regulation induces a seemingly high resolution in optical images because the near-field optical signal is strongly dependent on the tip–sample distance. To prevent the artifacts, imaging of point-like emitters (more ideally, buried beneath flat surface) have been recognized to be suitable.

Imaging of single nanoemitters such as dye molecules and quantum dots was of considerable interest from viewpoints of their fundamental physics as well as estimation of spatial resolution. However, the optical throughput (transmission efficiency) of aperture probes was too small to detect extremely weak signal from single nanoemitters. Owing to a sophisticated and simulation assisted design of aperture probe, 1% transmission efficiency of 100 nm aperture was achieved in the middle 1990s. Then, with the further progress in probe design and fabrication, aperture NSOM now performs imaging spectroscopy with 10–30 nm spatial resolution. Figure 4.2 shows a single molecule fluorescence image with a 10-nm resolution [2]. Currently such high performance probes are commercially available and contribute to expanding the range of NSOM users as a standard tool for nanospectroscopy.

With some delay from aperture NSOM, scattering NSOM was proposed in the early 1990s. It began to raise significant interest triggered by the demonstration of nanoRaman spectroscopy and nonlinear spectroscopy, thanks to the field enhancement effect. Further more interest will be attracted

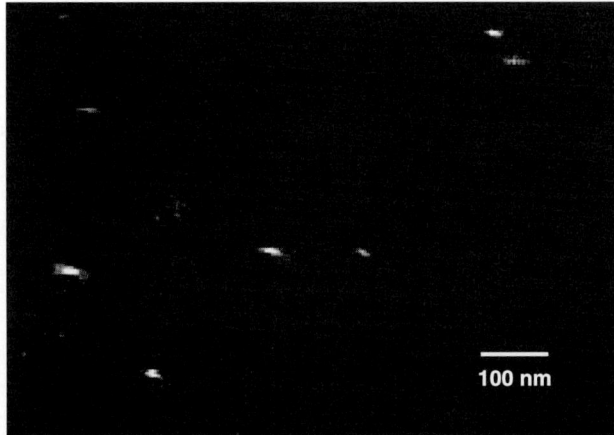

Fig. 4.2. Fluorescence image of single molecules measured by 10-nm aperture NSOM

by applying to the study of nonfluorescent materials like analysis of DNA molecules and diagnostics of silicon device.

4.3 Current State of NSOM

4.3.1 Probe

Most of aperture-type probes are fabricated based on the conventional optical fiber. After tapering of an optical fiber by means of chemical etching or the heating and pulling method, the tip is coated with opaque metal film. An aperture is created by milling the tip with focused ion beam [3] or, more easily, by pounding the tip against a hard substrate [4]. In any method, flatness of aperture face is essential to obtain the expected spatial resolution defined by the aperture size.

The efficiency in light transmission through the aperture decreases with aperture size naturally. By appropriate design of tapered waveguide structures or aperture shape, however, the degradation of throughput in light transmission can be minimized. It is theoretically and experimentally demonstrated C- and H-shaped apertures can provide substantial optical throughput [5]. A simple process to fabricate such complicatedly shaped aperture with micromechanical machining should be developed. As a more practical use, a triangular aperture is also promising to enhance both the spatial resolution and light transmission [6].

There are a few problems in use of fiber-based probes. Autofluorescence and Raman scattering from the fiber probe causes considerable background in weak signal detection, such as in case of single molecule imaging. Moreover unstable polarization state in transmission through the fiber degrades the sensitivity of polarization NSOM. A quick and effective solution to these problems is to make the fiber probe as short as possible. More drastically, a cantilevered aperture probe is advantageous to reduce the background and to maintain polarization state due to short transmission path length through the probe [7]. This also allows the regulation of tip–surface distance based on atomic force microscopy.

The scattering NSOM probe is much easier to fabricate compared with the aperture probes. In most cases, a cantilever coated with metal film is used. Silver is most suitable to obtain the electric field enhancement at the probe apex because the imaginary part of the dielectric constant is small. Preparation of a probe apex with a "hot spot," at which the electric field undergoes strong and stable enhancement, is important for reproducible measurements.

4.3.2 Operation Environment

In a variety of nanospectroscopic measurements of semiconductor materials, NSOM operation at low temperature and under an external field (magnetic, electric, or strain field) is required. There have been developed several types of low-temperature NSOM, where the entire scanning head is cooled in a liquid-helium flow cryostat [8] or only sample is cooled by attaching it to the cold finger in a vacuum chamber [9]. Since they have advantage and disadvantage, the design is optimized carefully taking account of operation mode and optical configuration.

Design and implementation of magnetic-field NSOM have been reported by several groups [10]. Magnetic-field with 5–10 T can be applied to a sample by using a superconducting magnet. Local application of electric field to a sample is easily conducted by using metallized tip as an electrode [11]. However, the strength and distribution of electric field is subject to the shape of the tip. The tip can also be used to induce a highly localized strain in the area by pushing the tip into the sample [12].

For the application to biological studies, NSOM operation in water is indispensable. Under liquid condition the Q value of tip vibration for the distance regulation based on the tuning fork drastically decreases due to the viscosity. The degradation of force sensitivity results in fatal damage to the sample. Compared with the straight fiber probe, a bent fiber probe with lower spring constant can reduce the tip–sample interaction force even in the contact-mode operation [13]. There is another idea originating from a diving bell, in which the tuning fork is vibrating in air, while the tip is immersed in liquid [14].

4.3.3 Near-Field Optical Spectroscopy

Combination of NSOM with conventional spectroscopic techniques has been carried out since the start-up phase of NSOM. Fluorescence spectroscopy is most popular because it is easy to obtain high contrast images with valuable information. In case of observation of fluorescent materials with a certain emission yield, imaging with a 10–30 nm spatial resolution within a reasonable measurement time has been reported [15].

Raman spectroscopy using scattering NSOM is attracting considerable interest. Owing to the electric field enhancement at the tip, Raman spectroscopy of single carbon nanotubes is realized as shown in Fig. 4.3 [16]. Combination with nonlinear spectroscopy (coherent antistokes Raman scattering spectroscopy) can enhance the resolution as high as 15 nm [17]. Further improvement of the spatial resolution will be expected by a complete clarification the mechanism of the field enhancement and moreover a chemical enhancement.

Polarization-contrast imaging is a powerful tool inherent in optical microscopy. There are two ways to obtain an image contrast in polarization NSOM. In most cases, a polarization state changes due to some optical anisotropy in a sample like birefringence or chirality. In the other case, the polarization change is induced by an asymmetric feature of the probe tip, which is generated intentionally or accidentally. The magnitude of polarization change depends on the refractive index of the sample below the tip and therefore we can visualize the sample structures not having any optical anisotropy.

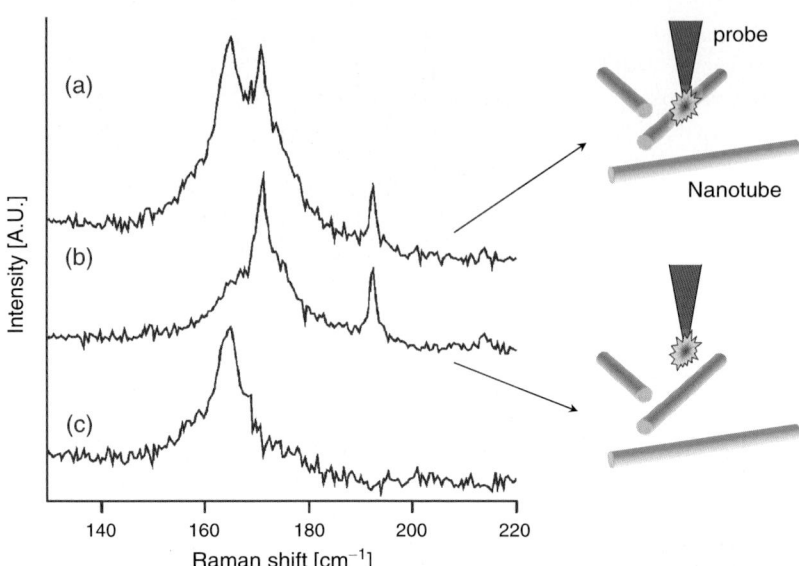

Fig. 4.3. (**a**) Near-field and (**b**) far-field Raman spectra of radial breathing mode of isolated single carbon nanotubes (**c**) is the subtracted Raman spectra between (**a**) and (**b**)

Figure 4.4 is an example of polarization aperture-NSOM employing the latter mechanism [18]. NiO nanochannel structure with a width of 20 nm and a depth of only 2 nm is clearly imaged in Fig. 4.4b. Because the flat end of the aperture does not trace the topographic feature on the surface as shown in Fig. 4.4c, the image is free from any artifact, i.e., the contrast and the resolution is purely optical.

4.4 Roadmap

4.4.1 Enhancement of Spatial Resolution

The spatial resolution in NSOM spectroscopy has already reached 10–30 nm. The reproducibility of such high resolutions is a key to offer standard nanospectroscopic measurements. Persistent efforts should be made to improve the probe fabrication, tip–sample distance regulation, stability of the scanning system and preparation of samples.

A resolution exceeding 10 nm will not be easy to achieve in a short term. In the case of aperture NSOM, the resolution is limited to 10 nm due to the finite penetration of light into the metal film. Higher resolution is likely to be attained in scattering NSOM. The trade-off between the resolution and the signal intensity will bring about a serious degradation of the image contrast. A solution for these problems is to utilize some additional physical and/or

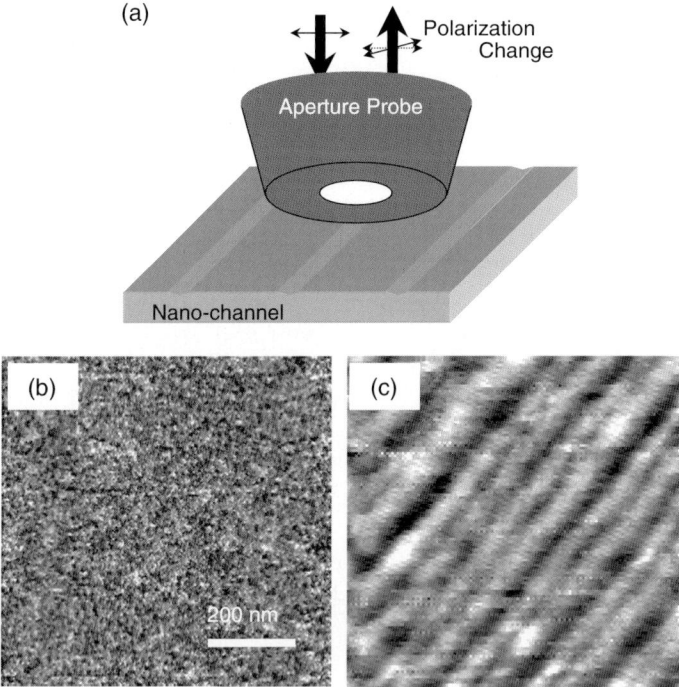

Fig. 4.4. (**a**) A schematic illustration of reflection NSOM imaging of an NiO nanochannel with a polarization contrast. (**b**) and (**c**) A topographic and an optical images, respectively

chemical mechanism. Optical detection of the charge transfer process due to the overlapping of electron wave functions of probe and sample will offer subnm resolution. A shift of sharp optical resonance induced by the application of local electric field or strain field is also promising. Although these are not general solutions, significant progress will be made by attractive applications with broad impact.

By enhancing the performance of NSOM in combination with advanced spectroscopic techniques, numerical simulation, and analysis technique, super-resolution imaging beyond the aperture size limit will be realized. Three-dimensional super-resolution spectroscopy will be used for a wide variety of applications including the observation of biomolecular dynamics and single dopant mapping in semiconductor devices.

4.4.2 Functional Probes

A progress in nanoprocessing will provide NSOM probes with some additional functions as well as the enhancement of spatial resolution, sensitivity, and mechanical stability. A metallized probe coated with an insulating polymer film except for its apex can be used as a probe of scanning electrochemical microscopy. The detection of physiologically active molecules in live cell imaging is an attractive application.

There have been many ideas on NSOM probes to give rise to the field enhancement by means of antenna structures like a bow–tie antenna [19]. This can be applied to control of the radiative process of nanoemitters, which has been experimentally demonstrated recently [20]. Emission from nanoemitters whose interaction with light is inherently weak becomes detectable through the enhancement of its interaction by appropriate design of antenna structure.

NSOM probes having a single molecule, quantum dot, or metal particle at the apex is indispensable to the fundamental study of their interaction with environment as a function of their precise positioning. Manipulation of nanoparticles and emitters without being destroyed, deformed, and bleached is strongly required.

4.4.3 Extension of NSOM Operation Wavelength

With a rapid progress in laser sources and photodetection devices, the wavelength range available in NSOM operation will be definitely more extended. Ultraviolet (UV) NSOM offers opportunities in the observation of nonlabeled biomolecules. Lasers and optical fibers available in deep UV have been already provided. Ten nanometer spatial resolution is critical because metal films conventionally used exhibit insufficient optical properties in UV region.

Combination of NSOM with near-infrared (NIR) spectroscopy is important in the investigation of materials for optical communications and the characterization of carbon nanotubes. High-performance photodetection devices based on InGaAs allows NIR spectroscopy with a sensitivity as high as in visible range. The optical throughput of aperture probe will drastically decrease with the wavelength. Careful design of NSOM probe to minimize the propagation loss in the probe is crucial.

Molecular recognition and imaging by means of midinfrared (MIR) vibrational spectroscopy has been widely used. For the combination with NSOM, the problem of optical throughput is more serious than in case of NIR. A cantilevered probe with a small aperture is proposed in place of the optical fiber-based probe [21]. Scattering NSOM is also promising to avoid the problem of small throughput [22]. Reduction of strong background from diffraction-limited area will be critical for high-resolution imaging.

Radiation at terahertz (THz) frequency is uniquely suited for broad range fundamental studies: high-speed semiconductor devices, superconducting materials and dynamics of biological molecules. THz microscopy based on the NSOM concept is now attracting significant interests. By minimizing a THz light source, high spatial resolution beyond the diffraction limit is directly obtained without using any kind of probe to localize illumination light. Development of efficient and monochromatic light source and high sensitive detection method will push forward a substantial progress in THz NSOM.

4.4.4 Nanoscale Light–Matter Interaction

An electron in a nanoscale confined system or in a regularly arranged molecule system exhibits a quantized energy level structure and the state is described by a wave function extended over the whole system. The system size is usually much smaller than the diffraction-limited spot of light and therefore we usually observe averaged optical response over the system. However, high resolution NSOM allows to look inside of the system: real-space mapping of electron wave functions. Figure 4.5 shows NSOM photoluminescence (PL) images of exciton and biexciton confined in a single GaAs quantum dot. The PL images reflect the shape of the exciton and biexciton wave functions and can be reproduced by a theoretical calculation [23].

Moreover, when a part of the system is locally excited by the NSOM probe, the interaction between light and the electron system drastically changes: selection rule on the electric dipole transition will be broken [24]. This makes a normally forbidden transition to be allowed and can be applied to novel photochemical processes [25]. By using the NSOM probe as an interface between the nanoscale electron system and the surrounding radiation field, we can control the storage, transfer, and release of optical excitation energy. This will give an idea to create nano-optical devices.

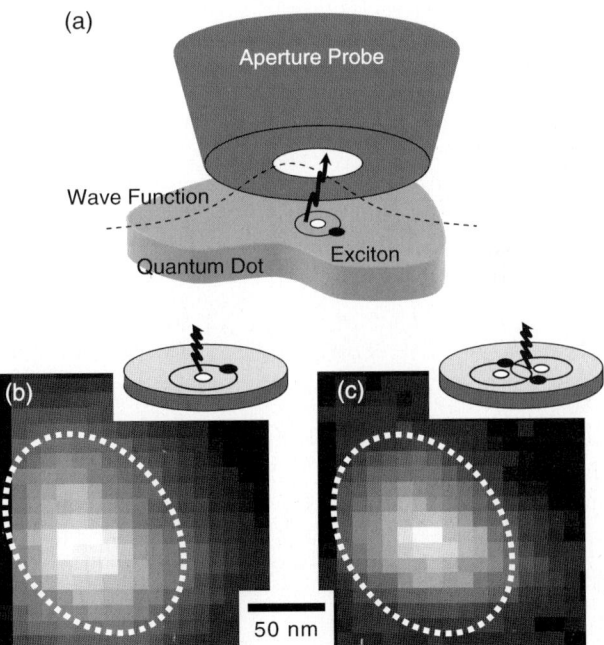

Fig. 4.5. (a) A schematic illustration of wave function mapping of an exciton confined in a quantum dot. (b) and (c) Spatial distributions of photoluminescence intensity from the single exciton state and the biexciton state, respectively

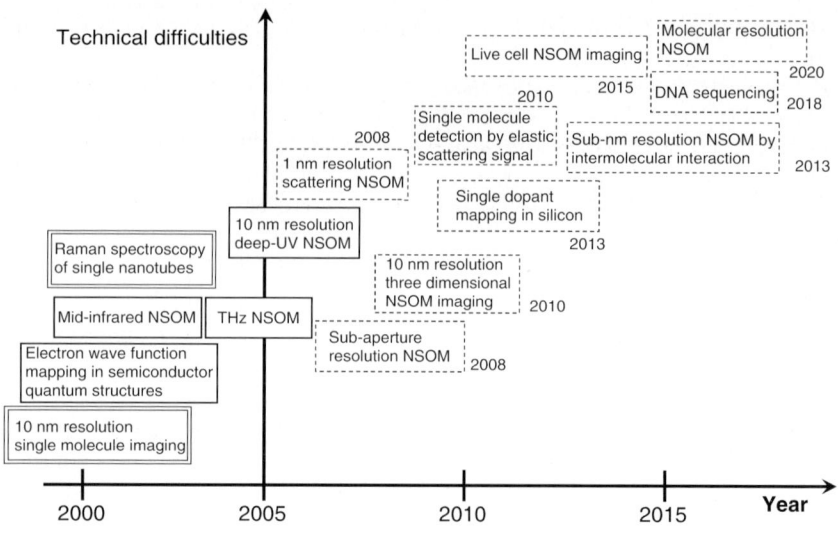

Fig. 4.6. Roadmap

As a summary of this section, Fig. 4.6 illustrates a roadmap for development and application of NSOM based on enhancement of the spatial resolution.

References

1. T. Saiki, in: *Near-Field Nano/Atom Optics and Technology*, ed. by M. Ohtsu (Springer, Berlin Heidelberg New York, 1998), pp. 15–29
2. N. Hosaka and T. Saiki, J. Microscopy **202**, 362 (2000)
3. J.A. Veerman, A. M. Otter, L. Kuipers, and N.F. van Hulst, Appl. Phys. Lett. **72**, 3115 (1998)
4. T. Saiki and K. Matsuda, Appl. Phys. Lett. **74**, 2773 (1999)
5. X. Shi, L. Hesselink, and R.L. Thornton, Opt. Lett. **28**, 1320 (2003)
6. G.C. des Francs, D. Molenda, U.C. Fischer, and A. Naber, Phys. Rev. B **72**, 165111 (2005)
7. R. Eckert, J.M. Freyland, H. Gersen, H. Heinzelmann, G. Schurmann, W. Noell, U. Staufer, and N.F. de Rooji, Appl. Phys. Lett. **77**, 3695 (2000)
8. T. Saiki, K. Nishi, and M. Ohtsu, Jpn J. Appl. Phys. **37**, 1638 (1998)
9. G. Behme, A. Richter, M. Suptitz, and Ch. Lienau, Rev. Sci. Instrum. **68**, 3458 (1997)
10. Y. Toda, S. Shinomori, K. Suzukim, and Y. Arakawa, Appl. Phys. Lett. **73**, 517 (1998)
11. T. Tadokoro, T. Saiki, and H. Toriumi, Jpn J. Appl. Phys. **41**, L152 (2002)
12. H.D. Robinson, M.G. Muller, B.B. Goldberg, and J.L. Merz, Appl. Phys. Lett. **72**, 2081 (1998)
13. H. Muramatsu, N. Chiba, N. Yamamoto, K. Homma, T. Ataka, M. Shigeno, H. Monobe, and M. Fujihira, Ultramicroscopy **71**, 73 (1998)

14. M. Koopman, B.I. de Bakker, M.F. Garcia-Parajo, and N.F. van Hulst, Appl. Phys. Lett. **83**, 5083 (2003)
15. K. Matsuda, T. Saiki, S. Nomura, M. Mihara, and Y. Aoyagi, Appl. Phys. Lett. **81**, 2291 (2002)
16. N. Hayazawa, T. Yano, H. Watanabe, Y. Inouye, and S. Kawata, Chem. Phys. Lett. **376**, 174 (2003)
17. T. Ichimura, N. Hayazawa, M. Hashimoto, Y. Inouye, and S. Kawata, Phys. Rev. Lett. **92**, 220801 (2004)
18. M. Sakai, S. Mononobe, A. Sasaki, M. Yoshimoto, and T. Saiki, Nanotechnology **15**, S362 (2004)
19. H. Sukeda, H. Saga, H. Nemoto, Y. Itou, C. Haginoya, and T. Matsumoto, IEEE Trans. Magn. **37**, 1234 (2001)
20. J.F. Farahani, D.W. Pohl, H.J. Eisler, and B. Hecht, Phys. Rev. Lett. **95**, 17402 (2005)
21. T. Masaki, Y. Inouye, and S. Kawata, Rev. Sci. Instrum. **75**, 3284 (2004)
22. B. Knoll and F. Keilmann, Nature **399**, 134 (1999)
23. K. Matsuda, T. Saiki, S. Nomura, M. Mihara, Y. Aoyagi, S. Nair, and T. Takagahara, Phys. Rev. Lett. **91**, 177401 (2003)
24. K. Cho, *Optical Response of Nanostructures: Microscopic Nonlocal Theory* (Springer, Berlin Heidelberg New York, 2003), pp. 121–128
25. T. Yatsui, T. Kawazoe, M. Ueda, Y. Yamamoto, M. Kourogi, and M. Ohtsu, Appl. Phys. Lett. **81**, 3651 (2002)

5

Scanning Capacitance Microscope

Yoshitsugu Nakagawa

5.1 Principle of SCM

Scanning capacitance microscope (SCM) is a variation of SPM which extracts the two-dimensional dopant profile in semiconductor devices by measuring the local capacitance with a tip [1]. When a metallic tip is placed on a semiconductor surface with an oxide layer in between, a metal–oxide–semiconductor (MOS) structure just like that in a semiconductor device is formed as shown in Fig. 5.1. SCM usually grounds the tip and a bias voltage (V) is applied to the sample instead of a gate electrode as in semiconductor devices. The dimension of the depletion region under the tip is detected by measuring the capacitance C between the tip and the sample and the voltage dependence of the capacitance (C–V characteristics) is mapped in the scanning area. As a practical matter, SCM measures a capacitance variation (ΔC) caused by a voltage modulation (ΔV) and obtain dC/dV because C is too small to be measured.

SCM signal is reviewed hereafter using a one-dimensional MOS structure model (Fig. 5.2) for simplicity. For a n-type semiconductor a negative

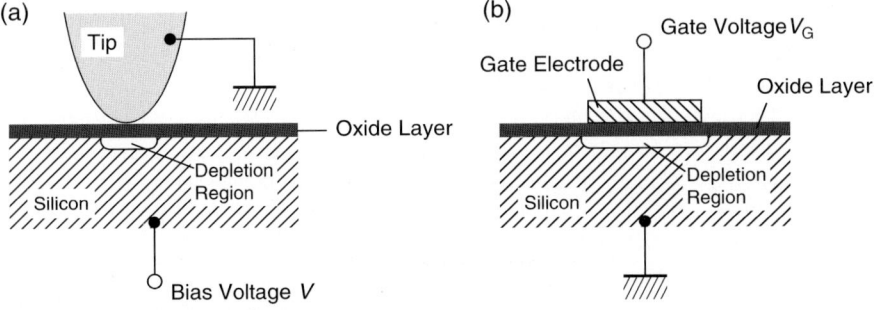

Fig. 5.1. (a) MOS structure of tip and sample in SCM observation scheme. (b) MOS structure in a semiconductor device

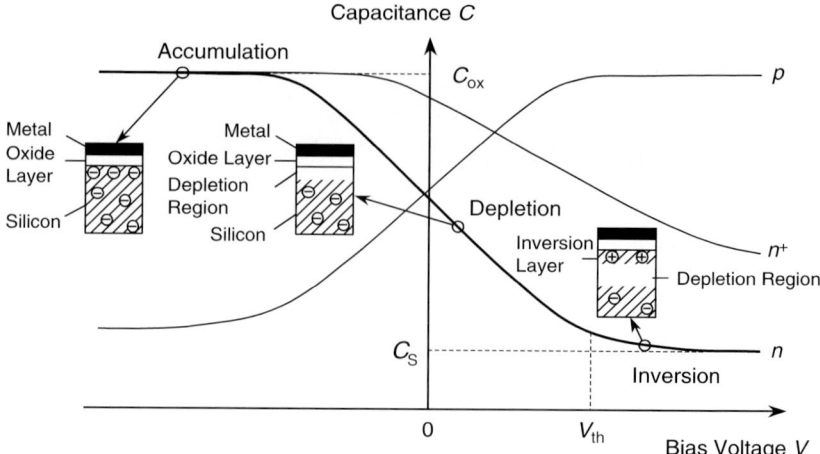

Fig. 5.2. C–V curves of one-dimensional MOS structure in SCM scheme

sample bias voltage causes majority carrier (electron) accumulation and then the MOS capacitance becomes equal to that of a capacitor formed by the oxide layer as a dielectric layer (C_{ox}). When a positive bias voltage is applied to the sample, a depletion region is induced just under the oxide layer and the MOS capacitance is decreased to the series connection of C_{ox} and the capacitance of the depletion layer. C decreases monotonically with positive V because of the increase of the thickness of the depletion layer. If V exceeds the threshold voltage V_{th}, the inversion occurs under the oxide and after that the C converges to a specific value of C_s because a higher bias voltage just accumulates the minority carrier (hole) in the inversion layer with a constant thickness (x_{dm}) of the depletion layer. (Here it is assumed that the number of the carrier in the inversion layer follows only the bias voltage, not the UHF voltage for the capacitance measurement. In reality the carrier in the inversion layer sometimes does not follow enough the AC bias voltage to make a deep depletion or it follows the UHF voltage due to minority carrier generation by photo excitation or minority carrier injection from 'p' layer closed to the measuring position.) Higher dopant concentration raises V_{th} and lowers x_{dm} (i.e., increase of C_s), resulting a lower dC/dV signal through a gentler slope in the C–V curve. In the case of p-type semiconductor, the majority carriers (holes) depletes at a negative sample bias, giving a dC/dV signal with the opposite polarity. Based on the principle mentioned above, we can distinguish the carrier type by the polarity of the dC/dV signal and evaluate the dopant concentration from the amplitude of the signal. In the real case a dC/dV signal has a peak at a specific dopant concentration and drops with decreasing dopant concentration lower than the peak value because of a shift in the flat band voltage and/or a non-negligible spreading resistance in series with the capacitance in measurement. So we need in the interpretation of an SCM

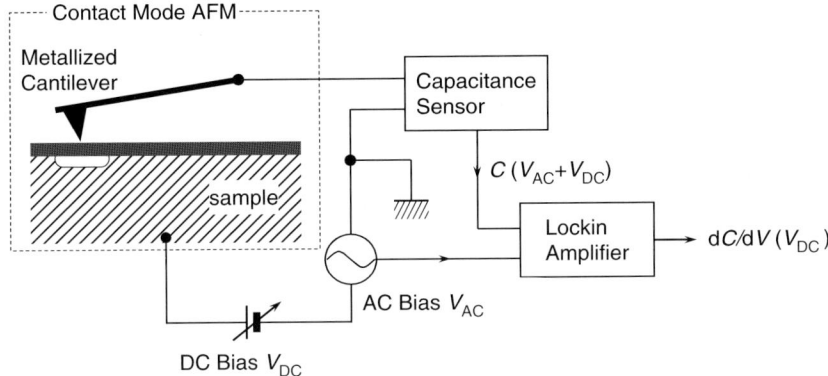

Fig. 5.3. Schematic illustration of SCM

image a enough care to the nonunique dopant concentration to the dC/dV signal obtained.

The instrumentation of SCM is a contact mode AFM combined with an electronic system that detects the dC/dV signal (Fig. 5.3). An RCA sensor for a video disc pickup is used for the capacitance detection. The sensor consists of a UHF (c.a. 1 GHz) oscillator, a resonator, and a detector. The capacitance between the sample and the tip is used as a part of the resonator. The capacitance variation conducts a resonance frequency shift of the resonator to change the amplitude of the UHF coming into the detector from the oscillator through the resonator. The dC/dV signal is obtained by lock-in detection of the detector output with an AC bias voltage (20–150 kHz) to the sample.

5.2 Practical Dopant Profiling by SCM

The needs for nanometer-scale two-dimensional dopant profiling are growing with the size reduction of semiconductor devices. SCM is, as a tool filling the needs, now widely used in structural evaluation and failure analysis of semiconductor devices.

Figure 5.4 shows an SCM image of an SRAM in a one-chip microcomputer as an example of the application of the microscope. The sample for the observation was prepared to have a thin surface oxide layer after exposing bare silicon surface by removing interconnects, electrodes, and oxides of the device. Figure 5.4a is an AFM height image in which a higher area is displayed in a lighter color. Figure 5.4b is a dC/dV image simultaneously obtained with the AFM image. In the dC/dV image a signal from p-type region is colored lighter and that from n-type region does darker, and the middle gray corresponds to the zero signal, corresponding to highly doped (metallic) region or depletion region along the pn junction. In the lower one-third of the image,

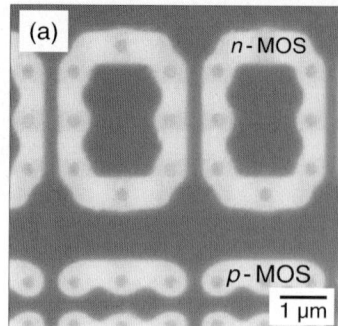 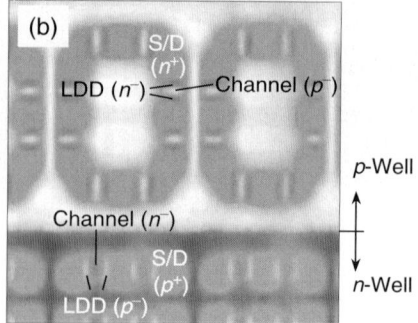

Fig. 5.4. (a) AFM height image and (b) simultaneously obtained dC/dV image of an SRAM in a commercial one-chip microcomputer

some p-channel MOSFETs are formed in an n-well and in the rest of the image some n-channel MOSFETs do in a p-well. An LDD gives a larger dC/dV signal than the source and the drain (S/D) and the channel shows the opposite signal polarity for both type MOSFETs. SCM has the capability to observe the dopant structure of a semiconductor device as shown in Fig. 5.4. In many cases cross-sectional structure has much importance and the cross-section of the sample along the evaluation line is prepared for the observation. An electrically isolated part in the observation area gives no signal and a bias voltage applied through a pn junction causes an overlap of the dC/dV signal from the junction, resulting a dC/dV signal inconsistent with the dopant concentration. To avoid these problems, it is useful to prepare the sample thin sliced and electrically contacted on the backside by metal evaporation.

The SCM observation is applicable not only to single-crystalline silicon but also to polycrystalline silicon and compound semiconductors such as SiC, SiGe, InP, GaAs, GaP, and GaN. Compound semiconductors usually do not have good oxide layer for insulation, but the Schottky barrier between the sample and the tip keeps them from an electrical contact. SCM can be also applied to an operating device by applying an appropriate voltage to each part during observation. The interpretation of the image thus obtained is not simple but in many cases the image can be interpreted as a reflection of the carrier profile of the device under operation. An example is given in Fig. 5.5 where an InP laser diode was observed with and without voltage application. The observation under laser emission, compared to that in a rest state, gives a higher signal from the active layers with a multiquantum-well (MQW) structure. This is because the carrier density at MQW is increased by the current passing through it. In Fig. 5.5b the upper clad layer seems to have an electrical contact with the current blocking layers on both left and right sides, and the electrons seem to flow into the blocking layers. In fact the diode has a failure of irregularly large lateral carrier diffusion.

The dopant concentration range measurable by SCM is 10^{15}–10^{20} cm^{-3} for single-crystalline silicon. The spatial resolution is determined by both tip

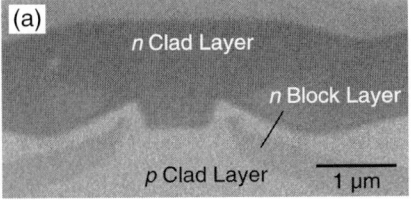 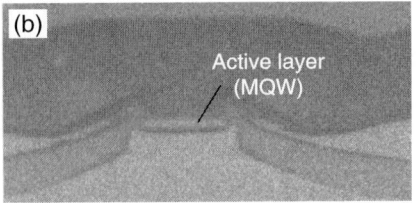

Fig. 5.5. SCM images of an InP laser diode. (**a**) Observed without external voltage, (**b**) observed under operation by an application of external voltage

radius and the expansion of the depletion region for dC/dV detection. The expansion of the depletion region is restricted by the detection limit of the capacitance sensor. In a commercially available SCM, the detection limit is 10^{-18}–10^{-19} F, which can detect ten dopant atoms or less. The sensitivity has already achieved an observation of irregularity in a pn junction contour due to unevenness in the dopant distribution in the scheme of discrete dopants.

One of the big problems of SCM is that the dC/dV signal and the junction line in consequence are considerably affected by the quality of the surface oxide layer (i.e., interfacial levels and fixed charges) and the electrical properties of the tip. Although the observation on the flat-band condition is desirable for sufficient reproducibility of the observation, quantitative evaluation of dopant concentration, and precise determination of the junction depth, the fulfillment of the condition all over the imaging area is very difficult be realized. At present, scanning spreading resistance microscope (SSRM) [2], a good alternative technique to SCM, is more suitable for these purposes as mentioned in Sect. 5.3.

5.3 Other SPMs for Dopant Profiling

In addition to SCM, STM/STS, SSRM, Kelvin force microscope (KFM), scanning nonlinear dielectric microscope (SNDM), scanning thermoelectric microscope, and a combination of dopant-selective etching and AFM topography are applied to two-dimensional dopant profiling of semiconductor devices.

SSRM, a kind of conductive AFM, is often used complementarily with SCM. It uses only one probe to measure spreading resistance and obtains a two-dimensional carrier density mapping. A typical SSRM can measure a wide range of carrier density (10^{14}–10^{20} cm^{-3}) using a current logarithmic amplifier with wide dynamic range. In comparison with SCM, SSRM has an advantage of the one-to-one (almost linear) correspondence of its signal to carrier density. The spatial resolution of SSRM is comparable or superior to that of SCM; the resolution of 10 nm has been already achieved. SSRM also surpasses SCM in the reproducibility in pn junction delineation. It, however, has some limitations: the carrier type cannot be distinguished by the image,

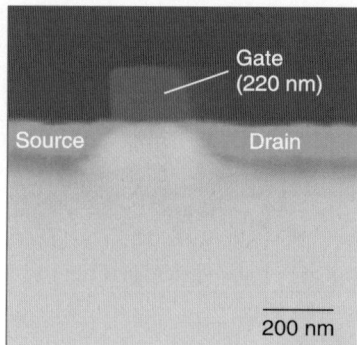

Fig. 5.6. SSRM image of an MOSFET with gate length of 220 nm

and SSRM destroys the sample surface during observations because of very high contact pressure. A complementary use of SCM and SSRM is effective to determine both carrier type and density.

Figure 5.6 shows an SSRM image of a cross-section of an MOSFET with a gate length of 220 nm. Compared to SCM, SSRM provides the more precise junction position and easier conversion of signal into carrier (dopant) density. It should be taken into account that different formula must be used for region of different carrier type to convert the resistance into carrier density because electrons and holes have different mobility. The conversion formula for single crystal cannot be applied to polysilicon gate also because of mobility difference.

For more precise determination of carrier density, a micro-four-point probe method has been developed as a variation of SPM [3]. The spatial resolution, however, is limited to several microns at present with 1.5 µm-pitched probes, being over 100 times larger than SCM or SSRM.

5.4 Roadmap

Higher resolution, more accurate dopant-concentration quantification, more precise junction delineation, and three-dimensional capability should be required for future dopant profiling of semiconductor devices. Taking these requirements and possible future achievements into account a roadmap of SCM is proposed in Fig. 5.7. The resolution of SCM can be raised by tip improvement and higher sensitivity of capacitance sensor. The tip resolution is determined in practice by a compromise with tip wear during scan. The improvement in the AFM feedback system may enable practical use of sharper tips by reducing the tip wear. As for the improvement of capacitance sensitivity, suppression of stray capacitance is effective. A shielded probe is proposed as an idea for this purpose [4]. Also effective to the increase in the capacitance

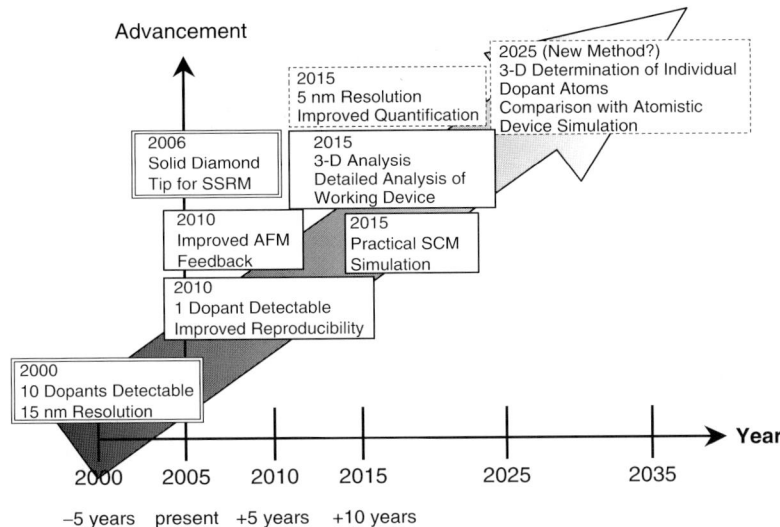

Fig. 5.7. Roadmap of SCM

sensitivity are to use higher frequency to detect the capacitance and to adjust the capacitance-detection frequency so as to maximize the output of the detection circuit by widening the tunable frequency range of the oscillator [5].

The dimension of the depletion region for dC/dV detection will be diminished to the volume that contains almost one dopant atom. The dimension, however, would have a minimum limit that is not negligible unless the SCM uses the depletion region as a probe. The resolution beyond the limit should be achieved by restoration of dopant profile from SCM signal using a computational procedure based on simulation. A regression procedure for this purpose has been developed already [6].

For the improvement in quantitative and reproducible measurement and the precise determination of pn junction line, SSRM is more promising than SCM at this moment. The problems of SCM, however, will be solved by adjustment of flat band voltage at each position on the sample surface during observation. The adjustment has been realized by control of DC bias so that dC/dV gives maximum at each position in the image [7]. A stray light streaming into a semiconductor sample during SCM measurement has been found to cause irregular dC/dV characteristics by generating excess electron–hole pairs. Both the ambient light and the AFM laser for optical deflection are the problem. The effect of the stray light obstruction has been investigated [8].

For the realization of three-dimensional measurements it has been tried that three-dimensional dopant profile is reconstructed from a series of SCM images with different depletion depth (i.e., different detection depth). The depth is controlled by changing modulation voltage ΔV in the differential detection system [9]. When precise simulation of SCM becomes easily utilizable, the three-dimensional reconstruction will be a practical tool for evaluating semiconductor devices.

As for other SPMs for dopant profiling the resolution of SSRM has been improved using solid diamond tips [10], which is desired to be commercially available. The capacitance sensitivity of 10^{-22} F has been achieved by SNDM, which is expected to be applied to high-sensitive dopant profiling.

With the shrinkage of semiconductor devices the perception of "continuous" dopant concentration becomes insufficient to explain the property of a device; the positions of individual dopant atoms should be taken into account [11]. When the future improvements of SCM in spatial resolution and capacitance sensitivity enable the evaluation of atomistic dopant profiles, it must be an important tool for the development of the semiconductor devices of the next generation.

References

1. J.J. Kopanski, J.F. Marchiando, and J.R. Lowney, J. Vac. Sci. Technol. B **14**, 242 (1996)
2. P. De Wolf et al., J. Vac. Sci. Technol. B **16**, 355 (1998)
3. T. Tanikawa et al., J. Surf. Sci. Nanotech. **1**, 50 (2003)
4. Š. Lányi, Ultramicroscopy **103**, 221 (2005)
5. J. Kwon et al., Ultramicroscopy **105**, 305 (2005)
6. J.F. Marchiando and J.J. Kopanski, J. Appl. Phys. **92**, 5798 (2002)
7. J. Yang and F.C.J. Kong, Appl. Phys. Lett. **81**, 4973 (2002)
8. G.H. Buh et al., J. Appl. Phys. **94**, 2680 (2003)
9. C.J. Kang et al., Appl. Phys. Lett. **71**, 1546 (1997)
10. D. Álvarez et al., Microelectron. Eng. **73-74**, 910 (2004)
11. A. Asenov, Nanotechnology **10**, 153 (1999)

6

Electrostatic Force Microscopy

Masakazu Nakamura and Hirofumi Yamada

Electrostatic force microscopy (EFM) is one of the SPM families which can visualize the distribution of electric potential, charge, contact potential difference etc. by detecting the electrostatic force between a probe and a sample. The first trial of the detection of electrostatic force is done by Martin et al. [1], and the observation of surface charges on polymer films was carried out by Stern and coworkers [2,3] which is thought to be the original work leading up to the current EFM technique. Kelvin probe force microscopy (KFM), which is analogous to a conventional Kelvin probe technique used to measure macroscopic contact potential differences, was proposed by Nonnenmacher et al. [4] By KFM, it became possible to quantitatively measure surface potential distributions. Scanning surface potential microscopy (SSPM) is basically the same as KFM, and EFM is occasionally used as the same meaning as KFM. Scanning Maxwell-stress Microscopy (SMM) [5] is a similar technique to measure surface potentials although method of measurement is slightly different.

In this section, fundamentals, present states, expected future and applications are explained for the EFM family including EFM, KFM, SMM, and related techniques.

6.1 Fundamentals

Let us consider the most simple system composed by a metallic probe and a flat metallic sample, of which Fermi energies are E_{F1} and E_{F2}, respectively. Assuming that their vacuum levels are equal, the Fermi levels are depicted as in Fig. 6.1. When both are electrically shortened to equalize Fermi levels, a work function difference V_S between two materials, which is described by the difference of work functions, forms an electric field between the two conductors. The source of this electric field is the charges induced on the surfaces of the probe and sample, which can be regarded as an electric double layer. If there exists a molecule in the gap, polarization of the molecule must be also considered.

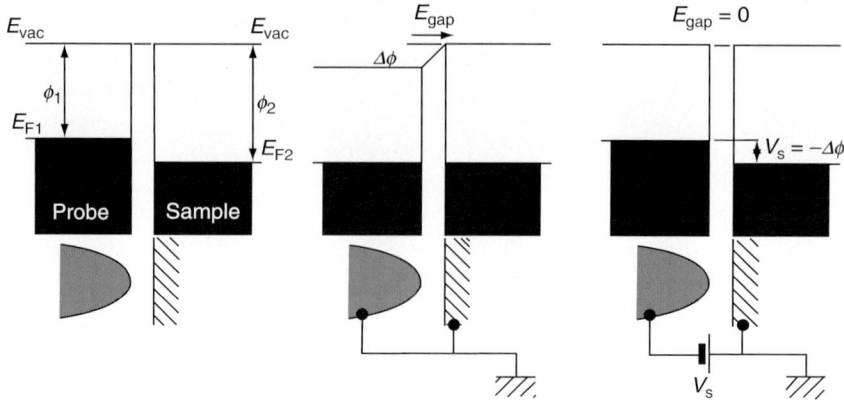

Fig. 6.1. Energy diagram of probe–sample system when (*left*) they are isolated, (*center*) both are grounded, and (*right*) vacuum levels are equalized by an external bias

When AC voltage V_{AC} with an angular frequency ω_m is applied between the probe and sample, the probe vibrates with the frequency ω_m by the electrostatic force. Using the capacitance between the probe and sample C, which is a function of probe–sample distance z, the electrostatic force F_z^{el} is described as

$$
\begin{aligned}
F_z^{el} &= -\frac{1}{2}\frac{\partial C}{\partial z}(V_S + V_{DC} + V_{AC}\cos\omega_m t)^2 \\
&= -\frac{1}{2}\frac{\partial C}{\partial z}\{(V_S + V_{DC})^2 + 2(V_S + V_{DC})V_{AC}\cos\omega_m t \\
&\quad + V_{AC}^2\cos^2\omega_m t\},
\end{aligned}
\tag{6.1}
$$

where V_{DC} is an externally applied bias voltage to the probe. Since electrostatic force is proportional to square of voltage, the vibration contains DC, ω_m and $2\omega_m$ components. The DC component is a static attractive force between the electrodes composing the capacitor, the ω_m component is a force to the previously described charges by the AC electric field, and the $2\omega_m$ component is a force induced to the capacitors only by the AC voltage. The ω_m component disappears when appropriate bias voltage is applied to the probe to cancel the contact potential difference (CPD), i.e., $V_S + V_{DC} = 0$. Surface potential or CPD can be therefore quantitatively measured by the feedback control of V_{DC} to maintain the ω_m component to be zero. In KFM operation, another interactive force between probe and sample, other than electrostatic force, must be detected to control probe–sample distance. There have been several methods classified by the combination of frequency modulation (FM) and amplitude modulation (AM) detections for these two different forces, which are listed in Table 6.1. Among them, an AM–AM method is widely used in

Table 6.1. Mechanical vibration and electric-field modulation methods in EFMs

Method	Control of probe–sample distance	Vibration frequency	Detection of electrostatic force	Electric-field modulation frequency
AM–AM	AM (tapping-mode)	$\sim f_{\text{res}}$	AM	Nonresonant
Lift-mode AM–AM	AM (tapping-mode)	$\sim f_{\text{res}}$	AM (lift-mode)	$\sim f_{\text{res}}$
FM–AM	FM	f_{res}	AM	$\sim f_{\text{res}}^{(2)}(\sim 6.3 f_{\text{res}})$
FM–FM	FM	f_{res}	FM	within FM bandwidth

f_{res}: fundamental resonant frequency of cantilever, $f_{\text{res}}^{(2)}$: second harmonic resonant frequency of cantilever, AM: slope (amplitude modulation) detection, and FM: frequency modulation detection

commercial instruments where the force for the distance control is measured by enforced mechanical vibration at the resonance frequency of the cantilever, namely slope detection, AM detection or *Tapping Mode*, and electrostatic force is measured by applying an AC voltage between probe and sample with a nonresonant, lower in many cases, frequency. However, this method has a disadvantage that the sensitivity of electrostatic force detection is low because it uses nonresonant frequency. Then, *lift-mode* was developed [6], where topographic and electrostatic signals are detected alternately. In lift-mode, topographic information is obtained by the first scan, and then the second scan without using a mechanical vibrator is performed to trace the topography by keeping a constant separation between probe and sample. During the second scan, the probe bias voltage is modulated by an AC voltage with the resonance frequency, which leads to a sensitive detection of electrostatic force. By lift-mode, cross-talk between topographic and electrostatic signals can be suppressed, although a drift of sample may cause an error in the potential measurement.

Detection of electrostatic force by noncontact-atomic force microscopy (NC-AFM) is done by choosing resonant frequency of cantilever as a modulation voltage to enhance the sensitivity, which corresponds to FM–AM or FM–FM method in Table 6.1. An FM–AM method is developed for KFM operation in vacuum where Q-factor of cantilever is too high to obtain practical response time of vibration amplitude. FM detection is used to maintain the probe–sample distance, and frequency of AC modulation voltage is set to the second harmonic of the cantilever to achieve high sensitivity and high-speed detection of electrostatic force [7]. In this method, electrostatic force is measured using AM detection. On the other hand, FM–FM method is operated by setting all the signals within the bandwidth of FM detection. Both topographic and electrostatic signals are detected by FM detection of the cantilever oscillation near the resonance frequency [8]. In this method, the electrostatic force is measured as a modulation of resonant frequency even if the modulation voltage applied to the probe is not in resonant. Accordingly, in FM–FM method, the electrostatic force induced by nonresonant ω_m causes the frequency shift of resonant frequency ω_0, which has an advantage of high voltage resolution.

6.2 Present State and Problems

Papers which use one of the EFM families as a primary tool and published from 1993 to 2005 (until June) were investigated to indicate the progress of the technique and vicissitudes of its applications. Figure 6.2 shows the graph categorized by physical quantity or phenomena to be measured, and Fig. 6.3 that categorized by the type of samples. Please note that not all published papers are included in this chart and that it is just an indication of a technical trend. Many papers reporting instrumental developments are in 1999 although such papers were rarely seen before that year. It is likely because atomic-scale resolution had begun to be expected for EFMs as NC-AFM exhibits technical progresses. This point is further discussed later. The progress and diffusion of NC-AFM also improved the sensitivity of electrostatic force detection, which resulted in extension of applications to wider ambient conditions and broader types of samples. For example, a result of SMM observation in a liquid ambient has been reported [9]. Since SMM does not use resonant vibration of cantilever, it is advantageous to use it in liquid where large damping makes the Q-value of cantilever small. Potential distributions in high-frequency operated integrated circuits were successfully measured by heterodyne detection [10,11].

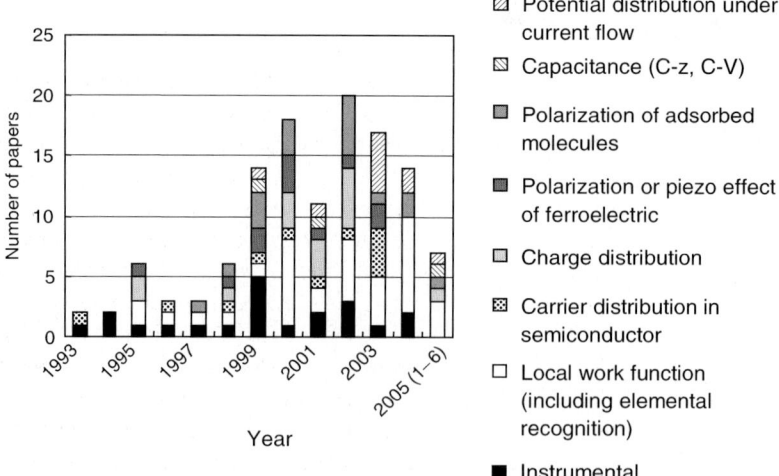

Fig. 6.2. Appearance of papers using any of the technique of EFM family as a primary tool: categorized by phenomenon of interest

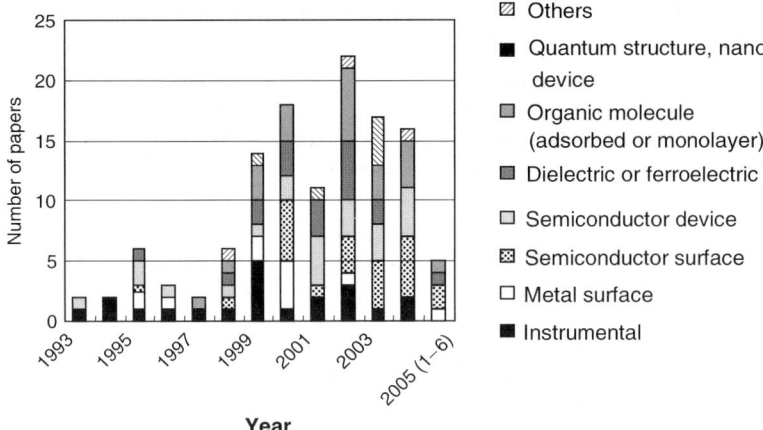

Others

■ Quantum structure, nano
  device

□ Organic molecule
  (adsorbed or monolayer)

■ Dielectric or ferroelectric

□ Semiconductor device

⊠ Semiconductor surface

□ Metal surface

■ Instrumental

Fig. 6.3. Appearance of papers using any of the technique of EFM family as a primary tool: categorized by sample to be measured

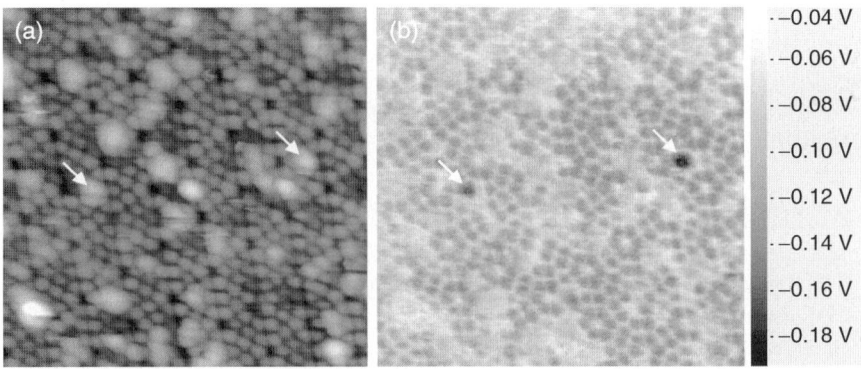

Fig. 6.4. $20 \times 20 \, \text{nm}^2$ (**a**) NC-AFM and (**b**) CPD images of Au adsorbed Si(111)-7×7

In heterodyne detection, high-frequency signal is added to the signal to be detected and the differential signal is measured to detect extremely high-speed periodical voltage oscillation.

Noncontact mode KFM with atomic resolution had evolved particularly in 1998 when the first international conference on NC-AFM was held. Kitamura et al. [12] reported atomic scale CPD images. Figure 6.4 shows (a) NC-AFM and (b) CPD images of a Au-adsorbed Si(111)-7×7 surface [13], which is obtained later as the work advanced. Au clusters adsorbed on the faulted half of 7×7-reconstructed structure is seen in the topographic image. Most of them exhibit higher CPD than that of surrounding 7×7 part. However, those indicated by arrows were found to have peculiarly lower CPD.

In such an atomic scale measurement using NC-AFM, the origin of detected force should be further discussed since it is impossible to completely separate electrostatic force from van der Waals or chemical forces [14]. Besides, an atomic-scale KFM image varies sensitively by the shape of probe tip as in the case of topography by NC-AFM [15]. How to obtain well-defined probes is therefore an important subject. Probe shape is essential, especially for EFM families because electrostatic force extends to the longest range among many interactions used in SPM, where not only the apex but also whole shape of the cantilever influences the result.

As for the instrumentation, KFM using a microfabricated cantilever with a piezoelectric deflection sensor was developed in 2002 [16]. Since photovoltaic effect disturbs the sensitive detection of surface potential on semiconductor surfaces, this type of nonoptical detection is a promising method for KFM.

There seems to exist only a few progresses in extraction method of more complicated information from raw data. For example, computerized tomography (CT) technique was reported in 1996 to obtain true charge distribution from EFM images [17], and experiments to measure density of states of a one-dimensional quantum wire or a two-dimensional semiconductor quantum well were reported in 1999 [18]. However, no similar work has followed up to this day. Further progress of this type of works could be earnestly desired.

On the contrary to the observation of vacuum level distribution by KFM, a complementary technique to directly observe Fermi level has been examined. It is because KFM has following problems when the electrical potential distribution in a working semiconductor device, which is a vital information for device physics, is of interest:

1. Work function which derived from vacuum level is not necessary in this case because the charge carriers never come out to "vacuum." Even if a CPD image under no bias voltages is subtracted as a background of KFM images, errors frequently remain.
2. Electrostatic force detected by a conductive probe is proportional to $\mathrm{d}C/\mathrm{d}z$ as the probe vibrates along z direction. When the surface under measurement has an abrupt shape, $\mathrm{d}C/\mathrm{d}z$ become complicated, and the CPD image measured tends to include a false contrast.
3. KFM readily detects the electrostatic forces not only from the part just below the probe but also from surrounding parts. Therefore, when a potential distribution of a working semiconductor device is measured, a false potential often appears near a biased electrode. The influence sometimes leaches to millimeter range.

To solve these problems, AFM potentiometry (AFMP) where a high-input-impedance amplifier is connected to a conductive probe has been developed [19, 20]. Figure 6.5 shows a schematic of the instrument with an organic thin-film transistor as a sample. During the tapping operation of an AFM with a conductive cantilever, electrical potential image is measured simultaneously with topographic image. Measurement of potential means that the Fermi energies of the probe and sample is equalized to be an equilibrium state, as in the center scheme in Fig. 6.1, by the charge transfer through the contact

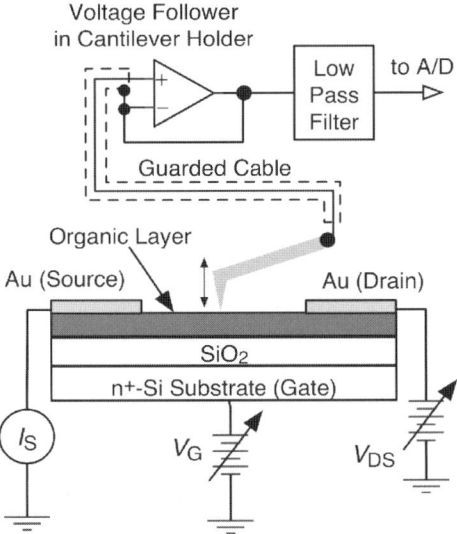

Fig. 6.5. Schematic diagram of AFM potentiometry measuring an organic thin-film transistor

point. This technique has an advantage that accurate potential is measured with high spatial resolution even near a biased electrode because the interaction takes place only at the contact point. Spatial resolution of 10 nm and potential resolution of 100 μV has been archived so far [21].

6.3 Roadmap

From the years around 2000, appearance of the papers in which any of EFM families is used as a major analytical technique increased. For physical quantity or phenomenon to be measured (Fig. 6.2), local work function measurement for elemental recognition on clean composite surfaces is one of the trends. Besides, the number of works studying polarization of submonolayer adsorbed molecules and potential distribution of working electronic devices are increasing. For samples to be measured (Fig. 6.3), semiconductor devices, dielectric/ferroelectric materials, and organic molecules/thin films are increasing in addition to the standard samples such as clean semiconductor surfaces. As mentioned earlier, the instrumental topic has been how to measure high-resolution images in gas or liquid ambient with high yields. Judging from these tendencies, expected progresses in a short term would be improvement and refinement as an analytical tool for small electronic devices and for local polarizations, and pursuit of higher spatial resolution in various ambient as an cutting edge research tool.

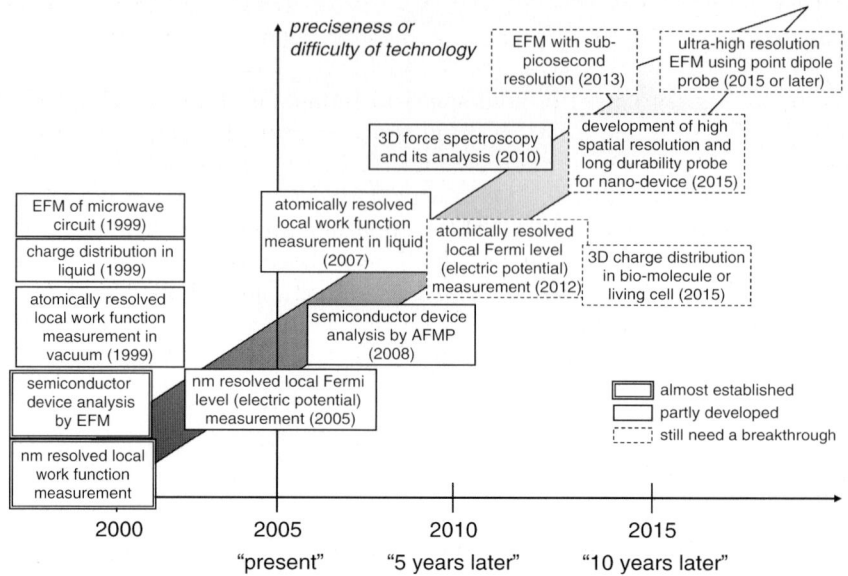

Fig. 6.6. Roadmap of EFM family

Taking these trends into account, the road map of EFM families was drawn up in Fig. 6.6 in expectation of further instrumental evolution. In the following part, detail of each technical subject is explained.

By considering the evolution speed of NC-AFM, atomically resolved KFM measurement would become possible in the near future even in liquid ambient. Pursue of higher resolution will go into the next phase attempting the extraction of electrical force from other forces such as chemical one. In vacuum, clear KFM images with atomic scale structures have been already obtained, in which interferences with topographic signal is confirmed to be negligible as much as possible. However, since dependence on probe–surface distance appears even in a chemical force [22], one must pay attention to such an interference especially when a high resolution KFM experiment is carried out by decreasing probe–sample distance. Ultimate prospect is that three-dimensional force spectroscopy, where force gradient is measured by sweeping probe–sample voltage at each three-dimensional matrix point, will be measured and analyzed in comparison with simulation to separate contributions by differently originated interactions.

As for AFMP, high spatial resolution is expected as far as the contact resistance is below its detection limit because the technique utilizes purely electrical interaction at contact point. For example, tunneling current is known to flow even in noncontact region if both the probe and the sample are conductive. Atomically resolved potentiometry by purely electric measurement will be realized if average tunnel resistance of around $10^9 \ \Omega$ during probe oscillation can be maintained. Combination of AFMP with KFM may be also established to visualize the band diagram of working microscopic semiconductor devices.

As for the measurement of high-speed phenomenon, high-speed sampling using pulse-width modulation has been already applied to measure voltage waveform with 10 ps-order time resolution [23]. Subpicosecond sampling of EFM signal would be achieved in the near future.

The CT technique to obtain three-dimensional charge distributions is also highly expected to be further developed. Used with an EFM instrument operated with liquid ambient, charge distribution in a bulky biomolecule or charge generation/transport in a living cell could be measured. For such a purpose, fabrication technique of conductive probes with well-regulated shape is essential.

References

1. Y. Martin, D.W. Abraham, H.K. Wickramasinghe, Appl. Phys. Lett. **52**, 1103 (1988)
2. J.E. Stern et al., Appl. Phys. Lett. **53**, 2717 (1988)
3. B.D. Terris et al., J. Vac. Sci. Technol. **A8**, 374 (1990)
4. M. Nonnenmacher, M.P. O'Boyle, H.K. Wickramasigh, Appl. Phys. Lett. **58**, 1921 (1991)
5. H. Yokoyama, K. Saito, T. Inoue, Mol. Electron. Bioelectron. **3**, 79 (1992)
6. H.O. Jacobs et al., Ultramicroscopy **69**, 239 (1997)
7. A. Kikukawa, S. Hosaka, R. Imura, Appl. Phys. Lett. **66**, 3510 (1995)
8. S. Kitamura, M. Iwatsuki, Appl. Phys. Lett. **72**, 3154 (1998)
9. Y. Hirata, F. Mizutani, H. Yokoyama, Surf. Int. Anal. **27**, 317 (1999)
10. G.E. Bridges et al., J. Vac. Sci. Technol. A **16**, 830 (1998)
11. V. Wittpahl et al., Microelectron. Reliab. **39**, 951 (1999)
12. S. Kitamura, K. Suzuki, M. Iwatsuki, Appl. Surf. Sci. **140**, 265 (1999)
13. S. Kitamura et al., Appl. Surf. Sci. **157**, 222 (2000)
14. T. Shiota, K. Nakayama, Jpn. J. Appl. Phys. Part 2 **40**, L986 (2001)
15. T. Shiota, K. Nakayama, Jap. J. Appl. Phys. Part 2 **41**, L1178 (2002)
16. N. Satoh et al., Appl. Surf. Sci. **188**, 425 (2002)
17. T. Ohta, Y. Sugawara, S. Morita, Jap. J. Appl. Phys. **35**, L1222 (1996)
18. D. Gekhtman et al., Mater. React. Soc. Proc. **545**, 345 (1999)
19. M. Nakamura et al., Synth. Metals **137**, 887 (2003)
20. M. Nakamura et al., Appl. Phys. Lett. **86**, 122112 (2005)
21. M. Nakamura et al., Proceedings of International Symposium on Super-Functionality Organic Devices, IPAP Conference Series **6**, 130 (2005)
22. T. Arai, M. Tomitori, Phys. Rev. Lett. **93**, 256101 (2004)
23. R.A. Said, J. Phys. D, Appl. Phys. **34**, L26 (2001)

7

Magnetic Force Microscope

Sumio Hosaka

7.1 Principle of MFM [1, 2]

Magnetic force microscope is for observation and measurement of stray magnetic field distribution by detecting small magnetic force or its force gradient using a fine magnetic probe based on an atomic force microscope (AFM). So far, the technology has been used as a magnetic balance, which is one of methods to measure macroscale magnetic field and magnetization with magnetic force. Magnetic force microscope (MFM) method has been developed for micromagnetic stray field image by using fine magnetic probe based on the magnetic balance and AFM. Features of MFM are (1) to detect very small magnetic force (10^{-13} N) or its force gradient (around 10^{-4} N m^{-1}) by using the small magnetic probe, which strongly works on the stray field emitted from a sample surface (magnetic domain) in a near-field range, and (2) to obtain the distribution image of the stray field magnetic domain with a resolution of less than 30 nm. In particular, it notes that the MFM image is made of stray magnetic field from the sample surface, but is quite different from magnetic domain image of an inner of the sample, which is observed by spin polarized-scanning tunneling microscope (SP-STM), spin polarized scanning electron microscope (Spin SEM), etc.

7.1.1 Estimated Resolution of MFM

Figure 7.1 a shows a model of MFM probe and a point magnetization in sample with a gap between them of ΔS, magnetizations of m and M in the probe and the sample, respectively, for the estimation of magnetic force and its force gradient against the probe. Assuming that the probe shape is conical, a magnetic force gradient $F'(Z)$ can be calculated by an integration of magnetic force gradient of small volume in a volume defined from the tip of the probe to distance of Z, from the tip with a parameter of the gap between the tip and the sample. The $F'(Z)$ is given by following equation:

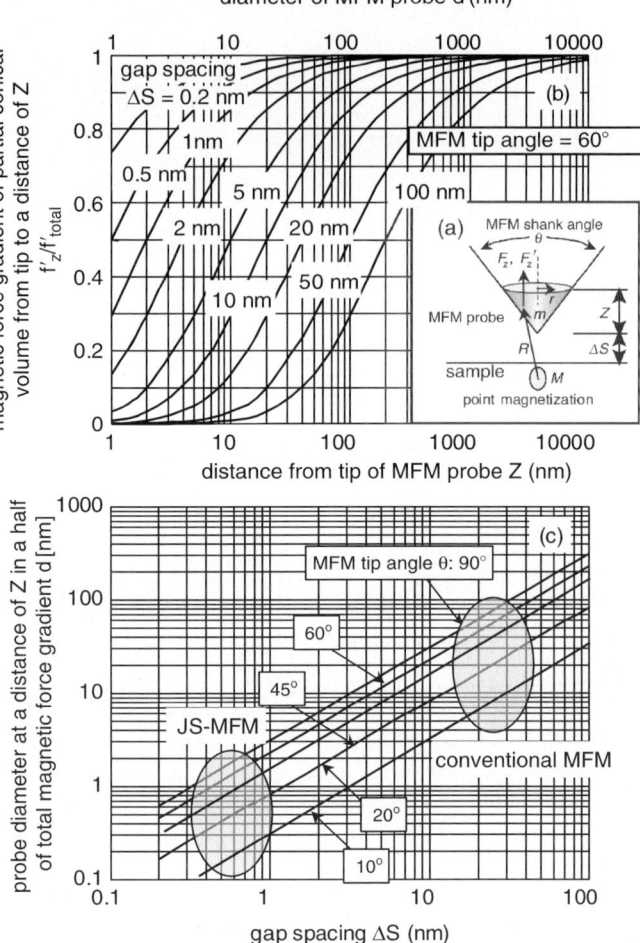

Fig. 7.1. MFM probe and point magnetization model (**a**), the estimated magnetic force gradient for distance of Z from tip of conical probe (**b**), and a change of the estimated probe diameter for gap spacing between tip and sample surface (**c**)

$$F'(Z) = -m \iint \frac{M(Z + \Delta S)r}{R^3} \mathrm{d}r\mathrm{d}\phi, \tag{7.1}$$

where r is the radial distance from a center axis of the probe. Using the equation, we can calculate the gradient $F'(Z)$, which a partial probe defined from the tip to the distance of Z receives in various gaps. We can define a diameter of MFM probe as a probe diameter at a distance of Z where the gradient becomes a half of the total force gradient. From Fig. 7.1b, effective MFM probe diameters can be estimated as shown in Fig. 7.1c. Figure shows that the MFM probe diameter becomes small as the probe approaches to the sample surface. This means that MFM resolution is improved by approaching

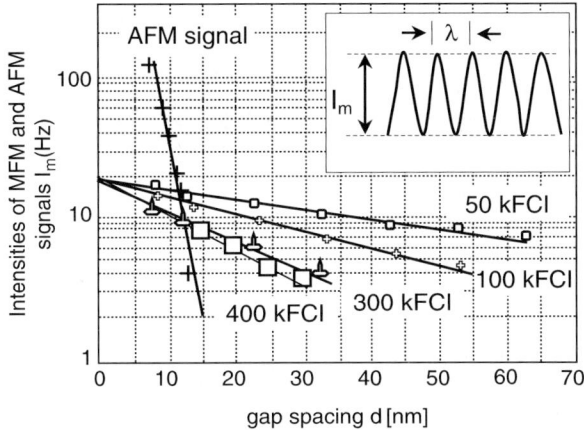

Fig. 7.2. Changes of atomic force gradient and magnetic force gradient signals due to gap spacing when detecting the magnetic recording pattern

the probe to the sample as close as possible. In such a near-field range, there are both atomic and magnetic force. Figure 7.2 shows both the atomic and the magnetic force gradient signals due to the gap spacing [3]. The magnetic force is one of the far field forces although the atomic force is one of the near field forces. This provides that the MFM signal is eliminated by atomic force signal in a gap within 15 nm. In practice, MFM image has to be measured by lifting the probe up from the sample surface with a gap of over 15 nm. There are two methods developed to lift it, one by using a noncontact driving under a constant force gradient control in applying electric force between the probe and the sample and the other by using a lift mode driving with a measurement of variable force gradient at a distance above the sample surface based on the surface data that is measured in contact or tapping mode driving. The latter is very useful and familiar with MFM users without a special technique.

When we observe magnetic recording media, their MFM images are different between in-plane and perpendicular recording medias. Figure 7.3 shows the typical MFM images. The image also changes due to a gap between the probe and the sample as same as the MFM resolution (Fig. 7.1). As we need to drive the probe at more than 15 nm above the sample surface without atomic force reaction, the image sometimes changes to poor one without the boundary information. The lifted distance is very important for weak magnetic field strayed from the surface when we measure high-density magnetic recording media (Fig. 7.2). We have to control the probe at around 15 nm in the lifting distance precisely.

7.1.2 Detectable Sensitivity of MFM

MFM's detected physical quantity is force gradient but not force. This is because a method to detect force gradient has about 3 orders magnitude of higher

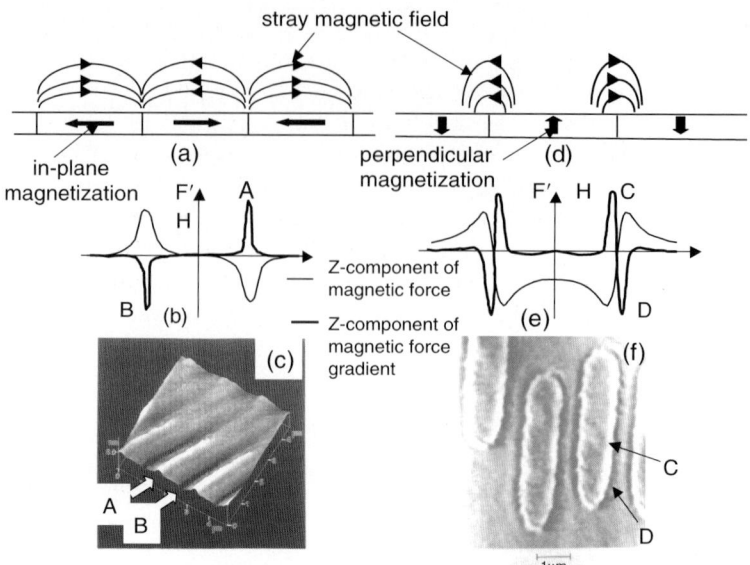

Fig. 7.3. Schematic diagram, Z-component of magnetic force and its gradient, MFM images; (**a**)–(**c**) in-plane magnetic recording and (**d**)–(**f**) perpendicular magnetic recording. ((**a**), (**d**): recording models; (**b**), (**e**): theoretical magnetic force and its gradient, and (**c**) and (**f**): MFM images of magnetic recording bit pattern with about 40 kFCI (flux change per inch), and magneto-optical recording bit pattern, respectively)

sensitivity than that to detect force. When we roughly transfer force gradient to force, MFM can detect weak force of about 10^{-12} N. This shows about 3 orders higher sensitivity than the force detection. The details will be described later. A method to detect the force gradient is used with optical beam deflection method same as that used in an AFM [4,5]. In a research phase, we sometimes used optical interferometer detection [6]. Today, cantilevers have a spring constant of 0.1 to several 10s N m^{-1} a resonant frequency of 10–150 kHz. They are available in commercial products.

Let us consider a magnetic force and its gradient of a magnetic moment working between magnetizations of m_1 and m_2 with a spacing of ΔS. A relationship between them is given by $F = -\Delta S F'$. On the other hand, minimum detectable magnetic force gradient is calculated with thermal vibration of the cantilever as follows:

$$F'_{\mathrm{m}} = \frac{1}{A}\sqrt{\frac{4k_0 k_{\mathrm{B}} T B}{\omega_0 Q}}, \tag{7.2}$$

where A is amplitude of the vibration, k_0 the spring constant, k_{B} the Boltzman constant, T the temperature, B the band of detectable frequency of detection circuits, ω_0 the resonant frequency, and Q is the quality factor of the cantilever.

The MFM has a potential to detect a magnetic force gradient of F'_m of about $10^{-4}\,\mathrm{N}\;\mathrm{m}^{-1}$. This corresponds to detectable force of $2\times10^{-12}\,\mathrm{N}$ when the gap is about 20 nm. Comparing the force with minimum detectable force in contact mode, the detectable force in noncontact mode is about 3 orders of magnitude higher than that in contact mode because practical minimum detectable force of 1 nN is considered. We have the experimental sensitivity of $5.0\,\mathrm{Oe}/10^{-10}\,\mathrm{N}$ [7] to transfer the magnetic field to magnetic force against the probe. Furthermore, the minimum detectable force gradient is improved by driving the system in a vacuum. The gradient in vacuum becomes small to a third to a tenth of that in air. Consequently, the estimated minimum detectable magnetic field is $5\text{--}16\times10^{-3}\,\mathrm{Oe}$.

7.2 History of MFM

MFM technology has been expected as one of the measurements and evaluations for micromagnetic structure and stray field as the magnetic recording density increases. In particular, the MFM needs becomes large in the field since around 1995. Since magnetic recording uses stray field, MFM has been applied to magnetic recording as a very suitable measurement for direct observation of the stray field near the recorded magnetic media surface with a gap of less than 100 nm. This microscopy has been developed and introduced by Martin et al. in 1987 [8], which is a year after the AFM has been developed by Binnig et al. The initial MFM probe was used with a sharpened iron or nickel wire. To obtain high resolution MFM image, the sputtered cobalt film on AFM cantilever and frequency modulating detection were used [9]. Then, large coercive magnetic film or soft magnetic film for the probe has been used. The MFM was used with optical interferometer for detecting cantilever position in early stage. As the detection has some technical issues in an alignment of the cantilever and linearity of the sensitivity, MFM with optical lever deflection detection has been developed and introduced by Hosaka et al. [10]. Furthermore, they have developed lift mode technique at each pixel [11]. Then, Digital Instruments Co. produced another type lift mode MFM in which the MFM signal is measured in every scan line. By some improvements described above, MFM technology becomes very important in measurement and evaluation of high density magnetic recording.

7.3 MFM Applications to Magnetic Recording Media

7.3.1 Observation of Ultrahigh Density Perpendicular Magnetic Recording

Figure 7.4a shows MFM image of random bit pattern perpendicular-recorded with a density of 52.5 Gb in^{-2} [12]. The track pitch is 280 nm, minimum bit length is 42 nm, which corresponds to 590 kBPI (bit per inch) in bit density.

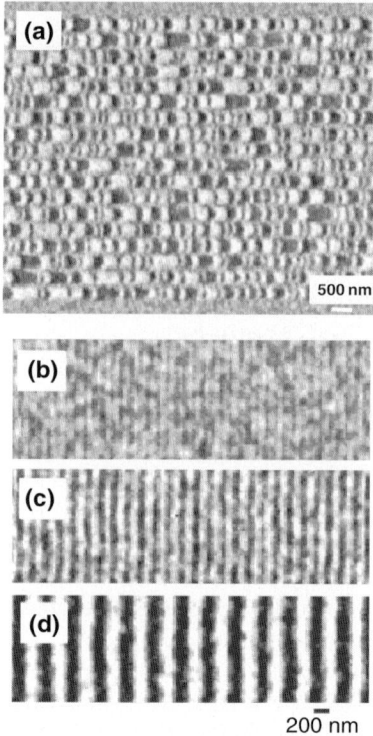

Fig. 7.4. MFM images of some perpendicular magnetic recording bit patterns; (a) 52.5 Gb in^{-2} RAM bit pattern [12], (b) with bit lengths of 42 nm (600 kFCI), (c) 63 nm (400 kFCI), and (d) 127 nm (200 kFCI) [14]

Figure 7.4b–d shows MFM images of high density magnetic recording bit patterns on double layer perpendicular magnetic media written by single magnetic pole head [13]. Figure 7.4b–d shows some bit patterns perpendicular-recorded with a bit density of 600, 400, and 200 kFCI, respectively. In Fig. 7.4b, a bit length of about 42 nm was observed. The experimental results indicate that MFM has a potential to observe fine distribution of stray field with a resolution of less than 30 nm.

7.3.2 Evaluation of Recording Property in High Density Magnetic Recording

We can estimate signal and noise property of the recording system including head and media from MFM images of the recorded media. Figure 7.5a shows a method to evaluate the signal–noise ratio of the media overwritten by the low or high frequency on the high or low frequency recorded area, respectively [14]. The recorded magnetic film was a perpendicular magnetic film of CoCrTa. The figure shows a scheme of simulating a detecting signal with a magnetic head, which has a width of magnetic resistance for reading, using

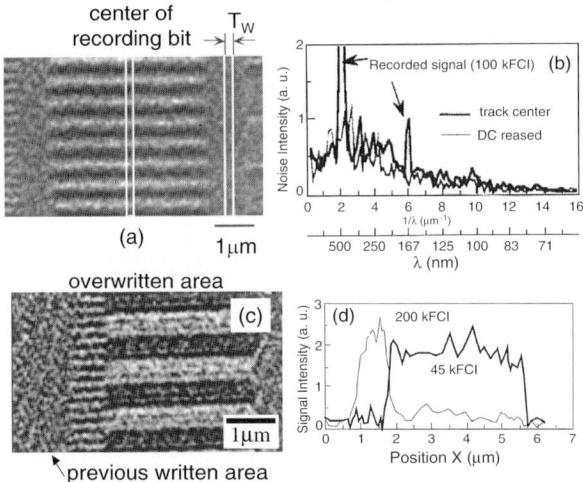

center of recording bit →|←T_w

(a)   1μm

overwritten area

(c)

1μm

↖previous written area

Fig. 7.5. MFM images and one-dimensional spectra analysis; (**a**) MFM image of perpendicular magnetic recording, (**b**) the spectra analysis using a narrow window with ΔT_m, (**c**) MFM image of off-track overwritten bit pattern with low frequency, and (**d**) intensity distributions of 200 and 45 kFCI signal components [15]

the MFM image. As shown in the figure, we can obtain and evaluate signal and noise components in the frequency spectra by Fourier transforming the rectangular area. Figure 7.5c and d shows overwriting property as described above. When we overwrite 45 kFCI signal on the magnetic media of CoCrTa film written by 200 kFCI, the MFM image was obtained as shown in Fig. 7.5c. In the figure, previous written magnetic bit pattern and overwritten bit pattern appear in a narrow area on the left side and a wide area on right side, respectively. We can observe the previous pattern remained in the overwritten area. When the MFM image was treated with Fourier transformation, the intensity distributions of 200 and 45 kFCI signals in Fig. 7.5c were obtained as shown in Fig. 7.5d. The experiments were performed with off-track overwriting method by 4.5 mm-wide magnetic head. The results indicate how the previous written signal remain there even when overwriting and erasing are performed. Figure 7.5d shows that the remaining signal component of 200 kFCI is in an area overwritten with a signal of 45 kFCI. This result indicates that there are some wrong parts of the head or the media in frequency property, magnetic field, magnetic domain size, etc. Thus, we can evaluate overwriting and erasing properties using the spectra and MFM images.

7.4 Roadmap of MFM

Mass storage capacity of magnetic and optical recordings has increased up at a rate of about 60% a year. In a recorded size of a high-end magnetic

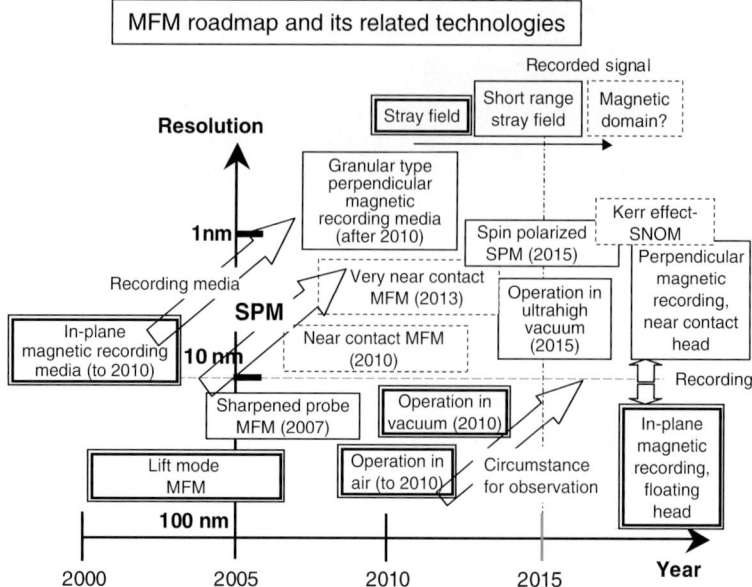

Fig. 7.6. Roadmap of metrological instrumentations for magnetic recording and trends of their related technologies

disc, its bit length is less than 50 nm, and a track length is around 250 nm, and driving gap spacing between the head and the disc is less than 20 nm in the storage system. Consequently, a high recording density was achieved with a density of higher than 100 Gb in^{-2} [15]. In the future, a recording system with a recording density of higher than 200 Gb in^{-2} and a spacing of around 10 nm will be achieved. As the capacity is glowing up, we need a microscope with a resolution less than 30 nm to observe very fine stray magnetic field at proximity on the film surface (Fig. 7.6). MFM technology is expected as one of the candidates to achieve such a need because it has some possibility to improve its resolution. In order to get higher resolution MFM image, we can consider that there are some ways such as (1) brash-up of MFM parts, (2) development of new type MFM technology in very small gap spacing, (3) use of very sharpened probe, and (4) operation in vacuum in stead of air.

On the other hand, MFM technology has been advanced by a development of magnetic recording technology. This means that a promoting force is given by a passion to develop high density magnetic recording. In this section, we consider a trend of magnetic recording, and then we will describe the MFM roadmap. Magnetic recording technology has developed an ultrahigh density recording media using a perpendicular magnetic recording. Next generation will be achieved by granular structure magnetic film or patterned media in the future. A size of the magnetic domain will be around 10 nm. Meanwhile, in read/write head, a single pole magnetic head or near-field optical head will appear. By these anticipations, MFM technology has to be advanced with higher resolution and sensitivity.

In higher resolution, MFM will be operated near a sample surface and use a sharpened MFM probe as shown in Fig. 7.1. The gap will be extremely small as same as contact mode. Furthermore, the sharpened probe will be used with an apex angle of less than $10°$. These technologies will provide a high resolution of around 10 nm in 2010. The former is just-on-surface JS-MFM, which has been introduced by Hosaka et al. [11]. The latter is nanotube probe, which will be used by 2010. In high sensitivity, we will use the vacuum operation, which will improve magnetic field sensitivity to 1 order higher of magnitude.

In the future, we will use magnetic domain in magnetic film for reading. In the case, we may have to modify the MFM technique by new concepts such as a spin polarized scanning probe microscopy (SP-SPM) or a near-field Kerr-effect optical microscopy, or an exchange force microscopy. Many MFMs and its related techniques will be available in micro- and nanomagnetic engineering.

References

1. S. Hosaka, A. Kikukawa, Y. Honda, Jpn. J. Appl. Phys. **33**, 3779 (1994)
2. S. Hosaka, A. Kikukawa, Y. Honda, Appl. Phys. Lett. **65**, 3407 (1994)
3. Y. Honda, Y. Hirayama, K. Ito, M. Futamoto, IEEE Magn. **34**, 1633 (1998)
4. G. Mayer, N.M. Amer, Appl. Phys. Lett. **53**, 1045 (1988)
5. S. Hosaka et al., Japanese patents no. 2138881 (1987.7.10) (in Japanese)
6. Y. Martin, C.C. Williams, H.K. Wickramasinghe, J. Appl. Phys. **61**, 4723 (1987)
7. S. Hosaka: IEEE Trans. J. Magn. Jpn. **8**, 226 (1993)
8. Y. Martin, H.K. Wickramasinghe, Appl. Phys. Lett. **50**, 1455 (1987)
9. T.R. Albrecht, P. Grütter, D. Hone, D. Ruger, J. Appl. Phys. **69**, 668 (1991)
10. S. Hosaka, A. Kikukawa, Y. Honda, H. Koyanagi, Jpn. J. Appl. Phys. **31**, L908 (1992)
11. S. Hosaka, A. Kikukawa, Y. Honada, H. Koyanagi, S. Tanaka, Jpn. J. Appl. Phys. **31**, L904 (1992)
12. H. Takano, Y. Nishida, M. Futamoto, H. Aoi, Y. Nakamura, Digest Intermagn., AD-06, Toronto, Canada, April, 2000
13. M. Futamoto, Y. Hirayama, Y. Honda, A. Kikukawa, in *Magnetic Storage Systems Beyond 2000*, NATO Science Series (Kluwer, Netherlands, 2001), pp. 103–116
14. Y. Honda, Y. Hirayama, K. Ito, M. Futamoto, IEEE Magn. **33**, 3076 (1997)
15. K. Yamamoto, Toshiba Reviews, **60**, 15 (2005) (in Japanese)

8

STM-Induced Photon Emission Spectroscopy

Tooru Murashita

8.1 Characteristics

Tunneling electrons injected from conductive probes can induce photoemissions from atoms, or from nanometer-sized structures of a sample. STM-induced photon emission enables us real-space characterizations of individual nanomatters including atoms, molecules, and nanometer-sized structures. Properties of the nanomatters are reflected in intensities, spectra, radiation-angle distributions, and polarities of emitted photons. In this section, "STM-induced" means "induced by tunnel currents." Photoluminescence (PL) obtained by photon excitation and cathodeluminescence (CL) [1] obtained by high-energy electron excitation have been commonly used for microscopic spectroscopy, but their spatial resolutions are not enough for characterizing individual nanomatters. In contrast, the STM-induced photon emission offers the ultimate spatial resolutions of nanometers or below. Tunneling electron beams are an excellent means of excitation for several reasons [2]:

1. The beam diameter is so fine that electrons can be injected into individual nanomatters.
2. The electron injection energy can be continuously and easily scanned or tuned in a wide energy range up to some 10 eV, which corresponds to the energy levels of most existing materials.
3. The size of the photon emission source can be reduced to nanometer levels because the mean-free-path of the electrons with such injection energies is as short as tens of nanometers.
4. High power density can be injected into nanomatters with low transmission losses.
5. Electron injection and extraction (i.e., hole injection) can be performed by reversing the polarity of the probe bias.

These characteristics make it possible to analyze luminescence from individual nanomatters and investigate energy-band structures of materials over a wide energy range. This enables us to distinguish properties of individual nanomatters that are hidden under statistical fluctuations in macroscopic measurements that detect emissions from a wide area. The STM-induced photon emission spectroscopy can be used to evaluate the electronic and optical properties of individual nanomatters even concentrated in high densities. In the early days, STM-induced photon emission spectroscopy was mainly applied to investigate semiconductor nanostructures and metal surfaces. Nowadays, it is also used to characterize organic dyes, biomolecules, fullerenes, carbon nanotubes (CNTs), and so on. Because electrons can be injected into an atom-sized area, we now see advanced applications such as the identification of individual atoms and molecules and the optical evaluation of the local structure of the individual polymers (such as biomolecules or the organic dyes).

Moreover, attempts are being made for the technique to apply to new technologies for the evaluation of the dynamic characteristics of photoemission devices, in the development of the ultra-high density optical storage, and for atom manipulations and nanofabrications by using it in combination with electronic and optical near-fields.

8.2 Emission Mechanism

STM-induced photon emission occurs when the electrons tunnel between the probe and sample. The emission mechanism, however, is not unique and is mainly electron–hole recombination radiation or surface plasmon emission. It is important to consider the difference in the photon emission mechanism when interpreting measurement results and applying the technology.

8.2.1 Electron–Hole Recombination Radiation

This kind of emission occurs when the tunneling electrons recombine with holes in direct-transition-type semiconductor and organic dyes as shown in Fig. 8.1a [3]. In p-type materials, injected electrons recombine with holes. In n-type materials, injected holes (removed electrons) recombine with electrons. Quasistable excitons are usually formed at the recombination. When the injection energy becomes as high as equal to or above twice the band gap, photoemission intensifies because the secondary electrons produced due to the impact ionization contribute to the recombination. Peak energy, full width half maximum and the strength of the emission sensitively reflect impurities, lattice defects, the quality of the material, and the atomic-layer-level size fluctuation of nanostructures. Moreover, information about the energy band structure can be obtained from the injection energy dependence (see Table 8.1 for a summary).

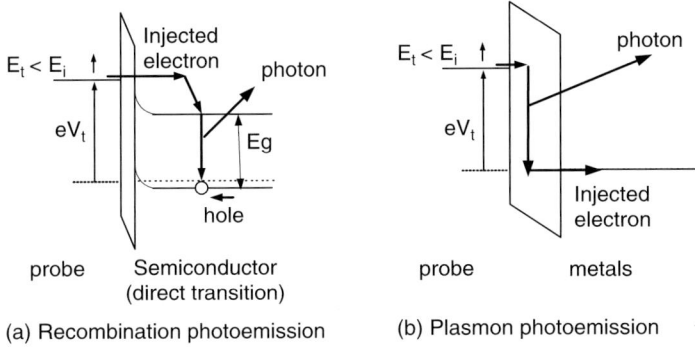

Fig. 8.1. STM-induced photoemission mechanism

Table 8.1. STM-induced photoemission spectroscopy

Measurement items	Evaluation subjects
Peak wavelength	Size of quantum structures
	Energy band structures
Peak width	Fluctuation in size of quantum structures
	Recombination speed (carrier lifetime)
Peak intensity	Occupation ratio
Specific distribution	Distribution of size of nanostructures
Injection energy	Internal quantum efficiency
dependence	(impact ionization)

8.2.2 Surface Plasmon Emission

For metallic matters, the electrons emitted from the probe to the sample relax during tunneling, and plasmons are excited by the differential energy as shown in Fig. 8.1b. Surface plasmon emission occurs when the translational symmetry of plasmons is destroyed by small structures or contaminants on a sample surface. Surface plasmon emission is stronger for gold and silver because the imaginary part of the permittivity is smaller in them and enhances electromagnetic fields. The emission spectrum is broad and the upper limit of the emission energy corresponds to the energy due to the probe-bias voltage. The properties in local regions of the nanomatters can be obtained by analyzing the strength, spectrum, and radiation angle of the emission. Surface plasmon photon emission is influenced by shape and material of the probe and the sample because plasmons do not diffuse, they localize in the narrow area between the probe and the sample. In this case, the radiation angle and the intensity of the emission sensitively reflect the refraction index and the single-layer level absorption and desorption on the sample surface. Information about the density of state distribution can be obtained from the electron energy dependence.

8.3 History of Research and Development

History of the research and development of STM-induced photon emission spectroscopy covering the period from 1988 (the year of the first observation) until 2000 is described in the 2000 roadmap [4]. The main highlights are follows. As for measured samples, photoemission image of atoms due to the surface plasmon emission from individual gold atoms was first observed in 1988 by Cooms et al. [5]. For semiconductors, STM-induced luminescence from a quantum well was observed (1990: Alvarado) [6], and then later from a quantum wire, quantum dot and fullerene. The spin of the tunneling electrons were investigated with optical polarization [7], and local plasmons were measured in high time resolutions [8]. Tunneling luminescence from organic dye was also detected [9].

As for equipment, on the other hand, the combination of a metal probe and condenser lens was employed initially [10]. The STM-induced photon emission is usually very feeble because the emission source is very small and the total injection power very low. Therefore, a parabolic mirror [3] and a fiber bunch close to the probe were introduced in order to make the solid angle as large as possible and to obtain high condensation efficiency while suppressing chromatic aberration [10]. In ultra-low temperature measurements, a large-diameter infrared-cut filter is placed on the cryostat to suppress heat irradiation from the outside [11].

Incidentally, the solid angle of the condenser is limited because the distance between the condenser and the photoemission source just under the probe apex is practically limited to the millimeter level.

As a way to improve luminescence collection efficiency, conductive transparent (CT) probes that inject electrons and simultaneously collect luminescence at the apex were developed, which is close to the photon emission source within nanometers [12]. Moreover, luminescence images of individual atoms doped into semiconductors can be obtained by coating thin film of silver on the apex of the CT probe to enhance surface plasmon emission [14].

8.4 Present Situation and Issues

Since 2000, recombination radiation from CdS nanoparticles (about $20\,\mathrm{nm}\phi$) (Fig. 8.2) [13] and surface plasmon emission from Ag nanoparticles have been reported. Also, there is applicability to three-dimensional structures like semiconductor nanowires because the photon emission from CNTs increases when CNTs detach from the substrate.

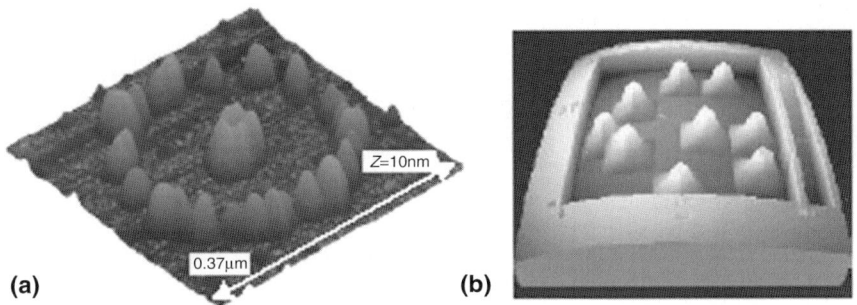

Fig. 8.2. Luminous semiconductor nanoparticles. (a) Topography image. (b) Luminescence image

As for the equipment performance, in future, improvements in the collection yields and the sensitivities of photon detection in order to advance to high-speed and high-quality measurements are important. The combination of the metallic probe with the lenses or mirrors is easy for use, but the lens has the problem that the aberration becomes bigger when the diameter becomes large.

At CT probes, plasmon photoemission enhancement has been obtained, but there is room for more optimization of the probe form and so on.

Probes for a shear-force mode AFM have been developed for application to samples with isolative regions, which are difficult to characterize by STM [15]. However, CT cantilevers for AFM will be needed in order to attain atom-size level spatial resolutions. For photon detectors, low-noise sensitivity for the photon counting method should be extended to infrared and ultraviolet regions. It is necessary to be able to reach a measuring point on a wide area of the sample in a short-time with atom-size-level high positioning accuracy. Combination with atom operation is also required.

8.5 Roadmap

Based on the present situation described in the last section, the following advances in the equipment performance and applications are expected within about the next 20 years.

8.5.1 Equipment Performance

The roadmap from the viewpoint of setup of the equipment is shown in Fig. 8.3a. The spatial resolution is currently high enough for practical use as evidenced by the success in obtaining luminescence images of individual-doped atom [14]. In the future, as the quantitative improvement of the spatial resolution reaches its limit, we will see a shift to improvement in qualities such as stability and clarity. The improvement of the collection efficiency and the extension of the detection wavelength will continue to be important.

(a)

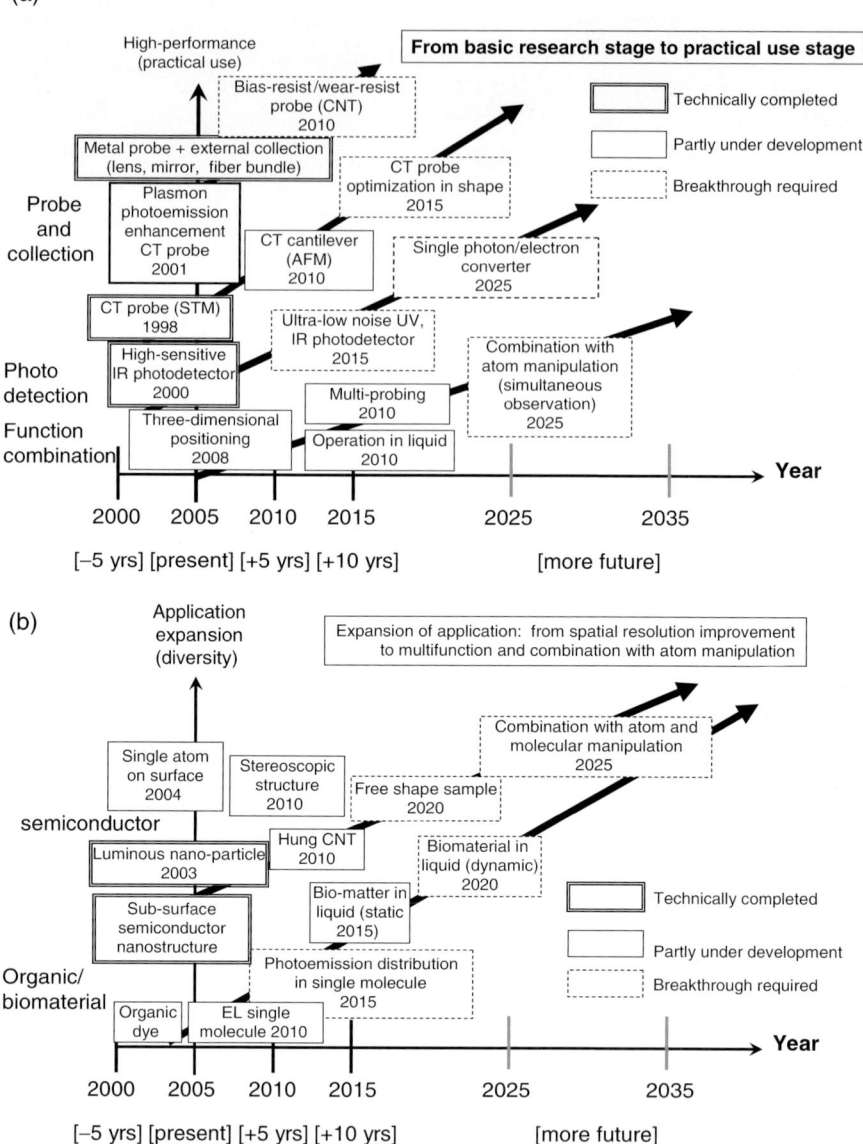

Fig. 8.3. (a) Roadmap of STM-induced photoemission spectroscopy (equipment performance: seeds). (b) Roadmap of STM photoemission spectroscopy (application: needs)

Collection Performance

In a setup where the current injection probe is separated from the lumines-cence collector, a CNT will be a suitable probe for stable injection of high current densities over a wide bias range because a CNT is attenuated and has high conductivity and superior wear resistance. At present, the size and fixation position cannot be controlled well. However, CNT probes will come into practical use within ten years. Luminescence collectors will be mainly the compact mirrors with low chromatic aberrations.

In the CT probe, on the other hand, the collection yield will be improved by enhancing surface-plasmon photoemission and optimizing the apex angle. And we see high durable CNT probes. These developments will occur within ten years.

For application to samples with isolative regions and to solid samples, conductive transparent AFM cantilevers will debut within five years and will become common within ten years.

Photodetectors that suppress background noise for faint light detection will be widely used in CCD and PMT within ten years, and the detectable wavelength will range from ultraviolet to infrared and will be applicable to the most kinds of samples. Also, within 20 years, we will see the developments of a photodetector capable of converting one photon to one electron (100% quantum efficiency).

Functional Combination

A CT probe that can inject current and collect light at only the probe apex will come into use comparatively early, which allow measurements of complicated biomolecules such as DNA in liquids. It will lead to advances in combination with SEM, three-dimensional probe positioning control with the atomic precision and unitizing the operation and measurement functions by combining the probes for atom operations.

8.5.2 Applications

The roadmap viewed from the standpoint of application fields is shown in Fig. 8.3b. In the future, STM-induced photon emission spectroscopy will be practically applied to the organic dyes and biomolecules as well as semiconductors.

Semiconductors

For about the next ten years, the main demand will be still on measurement of atoms and nanostructures. Within the next 15 years or sooner, we will see applications to samples with any property, including three-dimensional structures or insulating regions using AFM cantilevers.

Organic or Biomolecules

Within ten years, applications to investigating local parts of CNTs and organic EL molecules, which hopefully will serve as the core of the next generation of electronics, will be put into practical use. Within 20 years, it will be possible to analyze relationships between local structures of molecules and photoemission properties by manipulating atoms and molecular structures. Within 20 years, we will see STM-induced photon emission spectroscopy that will allow us to characterize or manipulate specific parts of biomolecules with the complicated three-dimensional structure such as DNA, by combining operations in liquid and atom manipulation.

References

1. B.G. Yaccobi, B. Holt, *Cathodoluminescence microscopy of inorganic solids* (Plenum, New York, 1990)
2. Ohtsu (ed.), T. Murashita, *Optical and electronic process of nano-matters* (Kluwer, Tokyo, 2001), pp. 181
3. E.O. Gobel, K. Ploog, Prog. Quant. Electron. **14**, 289 (1990)
4. S. Morita (ed.), *Scanning probe microscopes: Bases and future predictions* (Maruzen, Tokyo, 2000)
5. J.H. Coombs et al., J. Microsc. **152**, 325 (1988)
6. D.L. Abraham et al., Appl. Phys. Lett. **56**, 1564 (1990)
7. M. Pfister et al., Appl. Phys. Lett. **65**, 1168 (1994)
8. R. Berndt et al., Phys. Rev. Lett. **74**, 102 (1995)
9. S.F. Alvarado, Ph. Renaud, Phys. Rev. Lett. **68**, 1387 (1992)
10. R. Berndt, R.R. Schlittler, J.K. Gimzewski, J. Vac. Sci. Technol. **B9**, 573 (1991)
11. Y. Uehara et al., Appl. Phys. Lett. **76**, 2487 (2000)
12. T. Murashita, J. Vac. Sci. Technol. **B15**, 32 (1997)
13. K. Brabhakaran et al., Adv. Mater. **16**, 1495 (2004)
14. D. Fujita, K. Onishi, N. Niori, Microsc. Res. Tech. **64**, 403 (2004)
15. T. Murashita, Ultramicroscopy **106**, 146 (2006)

9

Scanning Atom Probe

Osamu Nishikawa

9.1 What is the Scanning Atom Probe?

A field ion microscope (FIM) [1] is the first instrument to realize the direct observation of individual surface atom and an atom probe (AP) [2] is a combined instrument of the FIM and a mass spectrometer with the sensitivity of detecting a single incoming atom and an ion. Accordingly, AP allows to mass analyze the surface atoms observed by the FIM. This unique capability is realized by utilizing the phenomenon called "field evaporation" [1]. When a high field is applied to surface, surface atoms are evaporated as positive ions. Since the field evaporation is an electrostatic process without any external energy, the evaporation takes place from the atom well exposed to the field, 20–$60\,\mathrm{V\,nm^{-1}}$, and proceeds in a highly ordered sequence, atom-by-atom and atomic layer-by-layer. Thus, the detection sequence of the evaporated atoms provides the depth profile of the specimen composition. This is the unique feature of the AP.

A specimen of the FIM is an apex of an extremely sharp tip in order to generate the high field required for the observation of FIM images and field evaporation of surface atom. The field evaporated ions are projected to the screen in the radial direction of the hemispherical tip apex and the magnification of the projected image is larger than one million. Since the position of the evaporated atom on the specimen surface and that of incoming position of the evaporated ion at the screen corresponds one-to-one, the detection of the ions incoming to the screen make possible to depict a two-dimensional map of composition at atomic resolution. Based on this idea, the screen is replaced with a position sensitive ion detector and the atomic layer-by-layer analysis visualizes the three-dimensional map of composition. This is the 3D-AP [3].

Although the unparalleled capability of the conventional AP and 3D-AP is highly attractive, the application of the AP has been severely limited because the fabrication of a sharp tip to generate the high field required for field evaporation is very hard for nonmetallic specimens such as diamond, organic molecules, polymers, and biomolecules. In order to overcome this difficulty a

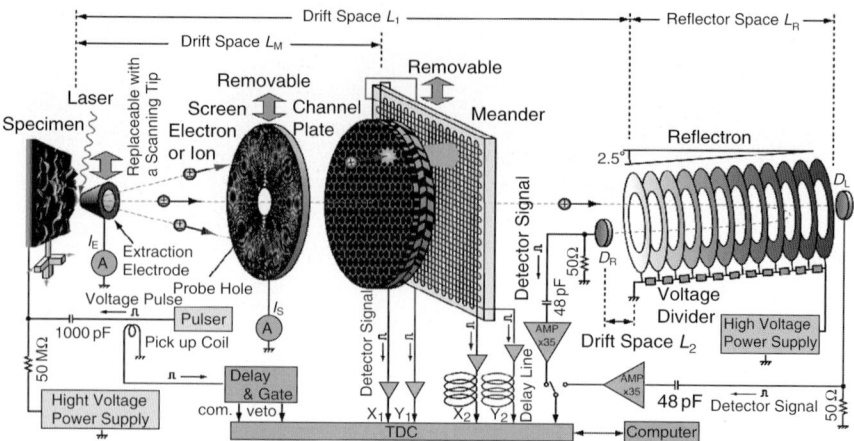

Fig. 9.1. Schematics of SAP with a position sensitive ion detector (Meander). For the high mass resolution analysis the Meander is removed to a side. Then the field evaporated ions go through the probe hole at the center of the screen and fly into the reflectron mass spectrometer and detected by the detector D_R. As a 3D-AP the screen is removed and the Meander is at the position behind the screen. For a low conductive specimen pulsed laser beam induces the photostimulated field evaporation

minute funnel-shaped extraction electrode is introduced. The electrode scans over a flat specimen surface with minute cusps. When the open hole at the apex of the electrode comes right above the apex of a cusp and a bias voltage is applied between the electrode and the specimen, the high field required for the field evaporation of specimen atoms is confined in a small space between the hole and the apex. This new type of the AP is named a scanning atom probe (SAP), Fig. 9.1 [4].

Atomic arrangement of a metal surface can be observed by the FIM of the AP. However, most nonmetallic specimens do not project a clear image on the screen. Accordingly the depiction of surface profile utilizing the scanning mechanism of the SAP is under progress. In this instrument a small cusp at the apex of the electrode serves as a scanning tip of a scanning tunneling microscope (STM). The combined instrument with an atomic force microscope (AFM) is also under examination.

9.2 Mass Analysis of Nonmetallic Specimens by the SAP

Since the introduction of the SAP, CVD diamond, graphite, carbon nanotubes [5], dissociation of organic molecules on titanium oxide and polythiophene are successfully investigated and the application area of the SAP is greatly expanded.

The examined polythiophene is a very soft thin film. The thickness of the film is approximately a few tens micrometers and it is utterly difficult to

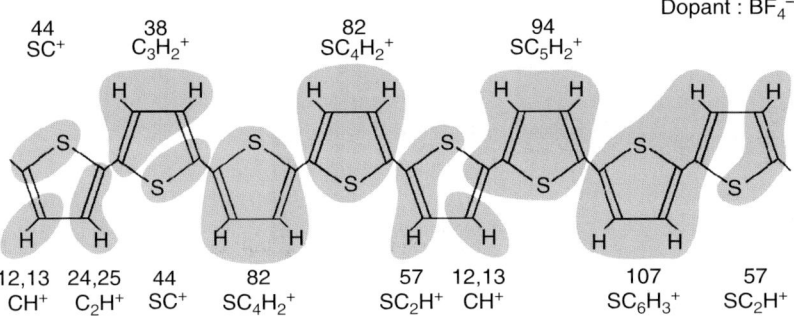

Fig. 9.2. Ideal structure of polythiophene. *Shaded areas* indicate the cluster ions

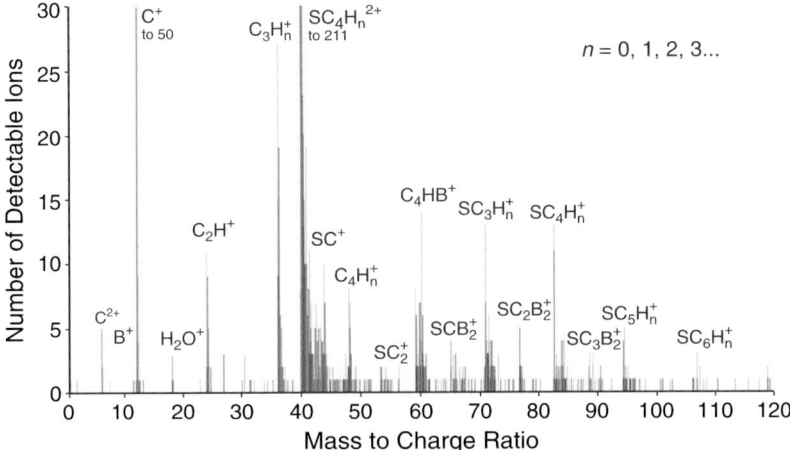

Fig. 9.3. Mass spectrum of polythiophene. Major mass peaks well agree with the fragment ions shown in the figure

make a sharp tip. Thus, the film was inserted between two nichrome sheets and cramped. Although no sharp tip is formed, microsharp corners of the film are sticking out. When the small open hole at the apex of the funnel-shaped electrode comes right above the sharp corner of the film, the field evaporation of polythiophene takes place.

Figure 9.2 is the ideal structure of polythiophene and Fig. 9.3 is the mass spectrum of the detected ions from the film. Interesting points are as follows:

1. Polythiophene is field evaporated as cluster ions.
2. The most abundant cluster is $SC_4Hn_2^+$, the radical of polythiophene. The doubly charged ions imply that the atoms forming the clusters are strongly bound each other.

3. No single sulfur ion is detected. This indicates that sulfur atoms are strongly bound with carbon atoms.
4. Most of the doped boron atoms are detected as a member of clusters.
5. Detection of few CH and SC cluster ions indicates the break of some double bonds. However, most double bonds are unbroken and stable.
6. When a negative bias voltage is applied, field emission current is measured. The variation of field emission current with applied negative bias voltage indicates that the polythiophene is semiconductive.

Fragmentation of organic molecules by the photocatalytic function of titanium oxide was also investigated. The masses of detected fragment ions are not randomly distributed but highly characteristics [6].

The present result indicates that the SAP has the capability of mass analyzing nonmetallic specimens and may suggest that the SAP open a new approach to investigate biomolecules at atomic level.

9.3 Present State and Problems

Basic function of the SAP is satisfactory working. However, the full automation of the scanning mechanism is not attained yet. Thus the combined instrument of the SAP and STM is still in the early stage of the development.

Fabrication of the extraction electrode is not automated yet. Optimization of the electrode shape is still in a trial and error stage.

The 3D mapping of composition by the conventional 3D-AP is advanced significantly in the last few years. Several tens million atoms are detected in 1 h and the 3D distribution of doped atoms in silicon is visualized. Since the size of single atom ions are quite similar, it is relatively easy to construct a 3D mapping. However, the construction of a 3D mapping of cluster ions requires a completely different treatment because the size and orientation of the clusters are not known. Accordingly, the development of 3D-SAP is not realized yet.

Mass resolution $m/\Delta m$ of the present 3D-AP is about 400 and that of the SAP is better than 1,000. However, this not high enough to mass analyze organic and biomolecules because the mass difference between CH_2 and N is only 0.0126 amu. Thus the mass resolution of more than 10,000 is necessary to discriminate CH_2 and N.

The detection efficiency of the ion detector is about 50–60% of the incoming ions because the channel plate detector has a dead area. In order to construct a perfect 3D compositional mapping, it is required to develop a new detector with 100% detection efficiency.

Lateral resolution of the position sensitive ion detector must be high. When the 100% detection efficiency and high lateral resolution are realized, the reconstruction of the structure of organic molecules is also realized.

Theoretical analysis of the experimental results is indispensable. Field evaporation of polythiophene is theoretically examined. The analysis supports the experimental result. The study may open a new method to investigate the binding state in the organic materials.

9.4 Roadmap

In order to estimate the future development of the SAP, six basic functions are examined. These are ion detection efficiency, mass resolution, lateral resolution, specimens, combination with STM and AFM, and theoretical analysis, Fig. 9.4

(a) *Ion detection efficiency.* The ion detector is the channel plate electron multiplier and about 50% of its incoming side is a dead area. In order to eliminate the dead area it is proposed to place a thin film in front of the channel plate. Such kind of research may improve the detection efficiency to nearly 100% by 2025.
(b) *Mass resolution.* The present mass resolution is higher than 1,000 and is sufficient for the mass analysis of metals and semiconductors. In order to mass analyze organic and biomolecules, the mass resolution of better than 10,000 is required. This mass resolution will be attained by 2020.
(c) *Lateral resolution.* The lateral resolution of the position sensitive ion detector shown in Fig. 9.1 is approximately 0.3 mm. This magnification is satisfactory to assign the position of individual detected ion on the

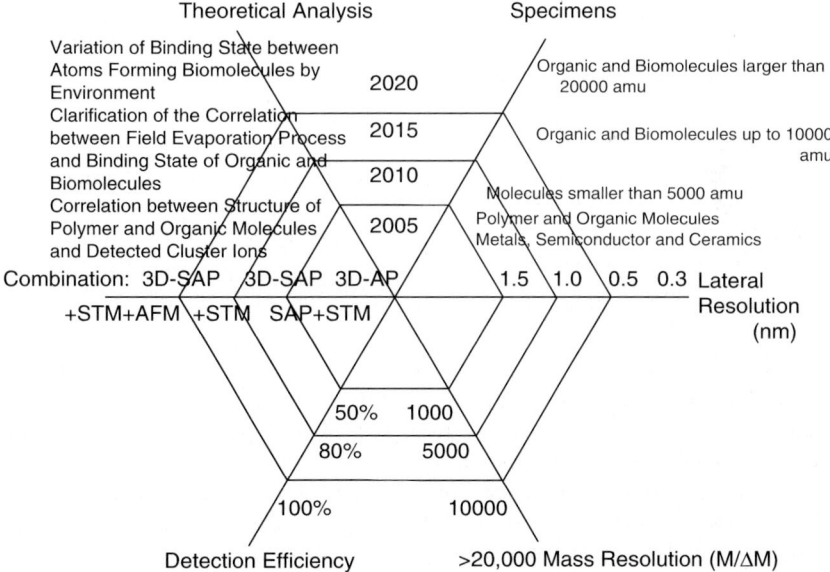

Fig. 9.4. Road map of six functions of SAP

specimen surface if the interatomic distance is about 0.3 nm and the magnification of the projected image on the screen is higher than one million. On the other hand the detection area of the detector is 40 mm square which corresponds to the area with 14,000 surface atoms. In order analyze a wide area the lateral resolution must be higher. The lateral resolution of the position sensitive ion detector will be improved to better than 0.1 mm by 2015.

(d) *3D mapping.* For metal and silicon specimens 3D mapping is in the satisfactory stage. However, 3D mapping of organic molecules and polymers are still remained untouched because the shape and direction of detected fragment ions are not known. It is expected that the mapping will be realized by 2020 or later.

(e) *Combination of instruments.* The first stage is the combination of 3D-AP and SAP. This combination is already at the final stage of the development. The second stage is to combine 3D-SAP and a high mass resolution AP. This is necessary to mass analyze organic molecules and biomolecules as mention in Sect. 9.3. The combined instrument of stages 1 and 2 is shown in Fig. 9.1. The third stage is to combine with an STM and AFM. The combination of the SAP and an STM is under progress and it will be realized by 2010. On the other hand the combination with AFM may take a longer time, possibly by 2020.

(f) *Theoretical analysis.* Theoretical analysis of field evaporation process has been thought fairly difficult. Unexpectedly, field evaporation of polythiophene was successfully analyzed in 2005 showing the agreement of the SAP analysis shown in Fig. 9.3. Although the analysis of organic molecules may be more complex than the polymer, it is expected to be analyzed by 2015 and the analysis of biomolecules will be attained by 2020.

References

1. E.W. Müller, T.T. Tsong, *Field Ion Microscopy, Principles and Applications,* (Elsevier, New York, 1969)
2. O. Nishikawa, K. Kurihara, M. Nachi, M. Konishi, M. Wada, Rev. Sci. Instrum. **52**, 810 (1981)
3. M.K. Miller, A. Cerezo, M.G. Hetherington, G.D.W. Smith, *Atom Probe Field Ion Microscopy,* (Oxford Scientific, Oxford, 1995)
4. O. Nishikawa, Y. Ohtani, K. Maeda, M. Watanabe, K. Tanaka, Mater. Charact. **44**, 29 (2000)
5. O. Nishikawa, M. Watanabe, T. Murakami, T. Yagyu, M. Taniguchi, New Diamond Front. Carbon Technol. **13**, 257 (2003)
6. O. Nishikawa, M. Taniguchi, S. Watanabe, A. Yamagishi, T. Sasaki, Jpn. J. Appl. Phys. **45**, 1892 (2006)

Chemical Discrimination of Atoms and Molecules

Tadahiro Komeda, Seizo Morita, and Yauhiro Sugawara

10.1 Recognition of Atom and Molecules; Inelastic Tunneling Spectroscopy

In the last volume of this series "SPM roadmap 2000," it was mentioned that the inelastic tunneling spectroscopy attracted attentions for a chemical identification of single-molecular level. In the last five years, there has been a large progress of this technique and it approaches to be a real tool for chemical analysis. Recent studies are briefly reviewed.

First let us summarize the mechanism of inelastic tunneling spectroscopy for the detection of vibrational modes by the use of tunneling current. The IETS is an all electron spectroscopy, thus it can be compared with the electron energy loss spectroscopy (EELS) which also detects the vibrational modes of adsorbates. However, the interaction between the incident electrons with adsorbate is through long-range interaction (dipole interaction). On the other hand, it is considered that the vibration excitation occurs with more proximity interaction between electrons and molecules for IETS [1]. Specially resonant tunneling mechanism is widely accepted for the excitation of vibrational mode. In the model, the tunneling electrons are first injected into the resonant level formed by the adsorbates, which stays in the level for a certain period of time and finally dissipated into the substrate. During the time period of the stay of electrons at the resonant level, the molecule is in ionic state and the nucleus distance tend to be changed, which is the driving force of the vibrational excitation. In the final state a vibrational mode (energy $\hbar\omega$) can be in an excited state if the energy of the tunneling electron exceeds $\hbar\omega$, which can be detected as a change in the conductance.

The IETS spectra are usually expressed as the second derivative of the tunneling current. For a normal peak, the feature should show a protruded shape $V > 0$ (V is the bias voltage applied to the substrate), and dip shape for $V < 0$. The conventional STS is measured mainly to detect the electronic state of the surface; thus the peak-width is relatively wide as $\sim 1\,\text{eV}$. On the other hand the vibrational features which the IETS measures have a sharp features whose width is in the order of ∼1 meV. Thus the lock-in amplifier should be used for its detection; numerical derivative of the I–V curve-like as used for STS measurement is no more appropriate. The lock-in amplifier is relatively slow measurement technique, which requires a long measurement time and the system should be in the feed-back loop off state. A difficult issue of this technique is how to keep the lateral and horizontal tip position during that time period. Currently the practical solution is to lower the sample and the head temperature to reduce the thermal drift [2].

The progress made in these five years can be summarized like following:

1. The instruments of Ho's group who showed a successful measurement of IETS is specially designed for this technique. Thus it was first considered that special instruments are required for its measurement. However, it was revealed that the stability obtained at 4 K is the most important issue for the technique, and IETS data are reported with the use of commercially available instruments [2, 3].

2. Successful measurements have been reported for simple gas molecule (acetylene, oxygen, and carbon monoxide), organic molecules (porphylene, phthalocyanine), and large molecules as fullerene and carbon nanotubes [2].

3. The measurement at above 77 K was reported for the IETS measurement on self-assembled monolayer. It is possible to obtain the IETS spectra even at room temperature, in principle, thus the development of the instruments might accelerate the measurement for wide range of the temperature.

4. Though theoretical reports have been continuously reported, it is hard to say that the selection rule and the excitation mechanism is established. For the latter issue, both the resonant excitation mechanism and the impact scattering mechanism are discussed.

5. If compared with the measurements of the conventional IETS, there are only limited number of reports of STM–IETS. However, there are many cases of unusual peak features reported in STM–IETS studies, which include negative peaks and Fano-shape peaks that have not been observed in the conventional IETS. This might be due to the strong interaction between the electronic states of the substrate and the adsorbate for the cases of STM–IETS which uses clean surfaces, while the molecules used in the conventional IETS are interacting with aluminum oxide. This indicates that case-dependent theoretical calculation is needed for STM–IETS, but also intriguing physical features can be expected.

6. Two-dimensional mapping of the strength of vibrational feature can give information both for basic understanding of the excitation mechanism and the future application of this technique for the search of functional group in a large size molecules. For the latter purpose, the rapid measurement capability is an important issue. Such an example is shown in Fig. 10.1c [3].

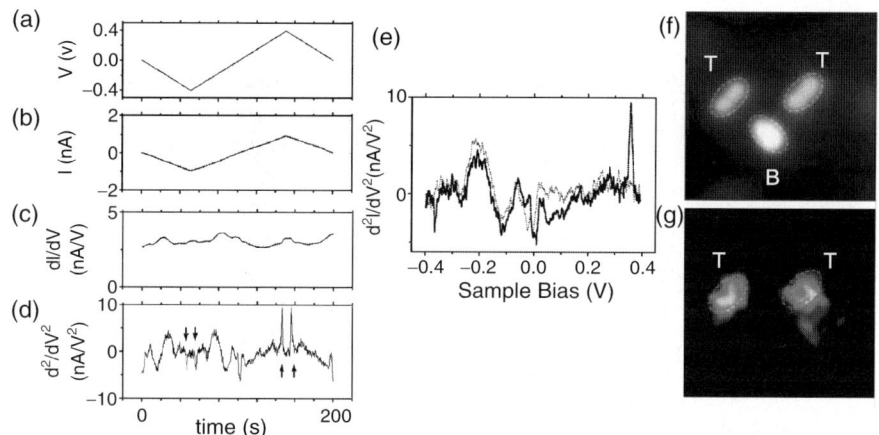

Fig. 10.1. Variation of sample bias (**a**), tunneling current (**b**) the output of lock-in amplifier of the harmonic (**c**), and the second harmonic component (**d**) as a function of the time in the measurement of the STM–IETS on a *trans*-2-butene molecule on the Pd(110). All graphs are the average of 16 cycles of the measurements (see text). The *arrows* indicate sharp features corresponding to the ν(C–H) vibrational mode. (**e**) The d^2I/dV^2 vs. V spectra obtained on a *trans*-2-butene molecule (*solid curve*) and on the bare Pd(110) surface (*dotted curve*). Simultaneous observation of the topographic image (**f**), and the mapping of d^2I/dV^2 intensity (**g**), on the surface which contains both of *trans*-2-butene (label T) and butadiene (label B) molecules. The tip scans the area with the feedback-loop on (tunneling conditions of $V_{\text{sample}} = 360\,\text{mV}$ and $I_{\text{tunnel}} = 1\,\text{nA}$, area $43 \times 43\,\text{Å}^2$)

In the top panel of topological STM image, we can see *trans*-2-butene molecule (T) and butadiene molecule (B) are co-adsorbed on Pd(110) surface. By observing the same area with the modulation voltage applied for the bias voltage and detecting vibrational feature with lock-in amplifier, the mapping of C–H stretching mode is obtained like as shown in the bottom pane. As the C–H stretching mode appears stronger for *trans*-2-butene than for butadiene, there are clear contrast for the two molecules in the mapping image showing the capability of chemical recognition.

10.2 Chemical Identification of Atoms by AFM

10.2.1 Chemical Identification of Atom Species by AFM

The first roadmap, "SPM Roadmap 2000" in Japanese, stated "At present, chemical identification of individual atoms and molecules by SPM is very difficult." As an example of chemical identification determined from electronic characteristics, it cited the chemical identification of Ga atom (empty state)

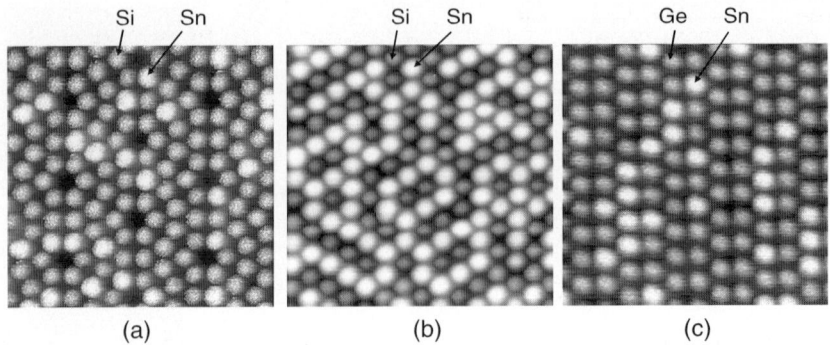

Fig. 10.2. NC-AFM topographies of intermixed surface (**a**) Sn/Si(111)7×7 [5], (**b**) Sn/Si(111)$\sqrt{3} \times \sqrt{3}$ [5], (**c**) Sn/Ge(111)-c(2×8) [6]

and As atom (filled state) of GaAs(110) cleaved surface, which were selectively imaged by the switching of polarity of bias voltage [4], that is, a kind of scanning tunneling spectroscopy (STS). At that time, there was no example of chemical identification of atom species by AFM. During the recent five years after "SPM Roadmap 2000," however, not only periodic atoms in the crystal such as GaAs(110) cleaved surface, but also randomly intermixed atoms consisting of heterogeneous atoms have been chemically identified not only by STS but also by AFM. Figure 10.2 shows noncontact-AFM (NC-AFM) topographies with chemical contrast (atom selective AFM imaging) on Sn/Si(111)7×7 [5], Sn/Si(111)$\sqrt{3} \times \sqrt{3}$ [5], Sn/Ge(111)-c(2×8) [6]. STS discriminates intermixed heterogeneous atom species with different local electronic density of states (LDOS) near the Fermi surface of free electrons by different magnitude of tunnel current or height difference, while NC-AFM discriminates intermixed heterogeneous semiconductor atom species with different covalent bonding potential and covalent bonding radius by different magnitude of frequency shift or height difference. STS indirectly depends on atom species because LDOS strongly depends on the crystal structure and also its sample size, while NC-AFM directly depends on atom species itself because NC-AFM measures short-range force such as covalent bonding force. Therefore, for the chemical identification of selected single atom, NC-AFM seems to be more suitable than STS. Chemical contrast of atom selective imaging with NC-AFM, however, strongly depends on the tip–surface distance. In general, as shown in Fig. 10.3, chemical contrast between intermixed heterogeneous atoms such as Si and Sn adatoms disappears by decreasing the tip–surface distance [7]. We can understand physical origin of such tip–surface distance dependence of chemical contrast from tip–surface distance dependence of frequency shift of the selected Si and Sn adatoms indicated by white arrows in the inset NC-AFM image as shown in Fig. 10.4a [7] (hereafter, we call frequency shift curve). By using the inversion procedure, we transformed frequency shift curves into force curves and obtained site-specific force curves

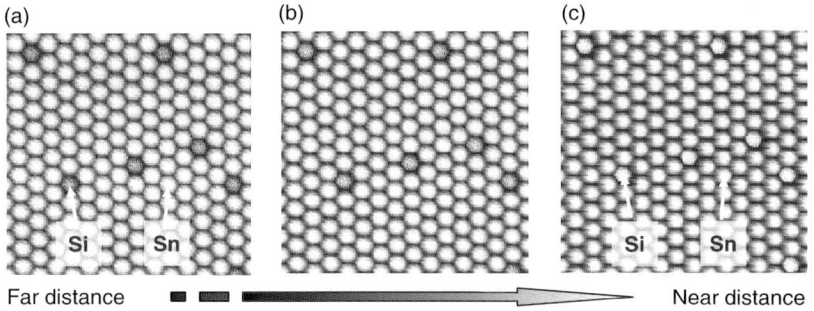

Fig. 10.3. NC-AFM topographies of intermixed Sn/Si(111)$\sqrt{3} \times \sqrt{3}$ pure phase surface as a function of tip–surface distance [7]

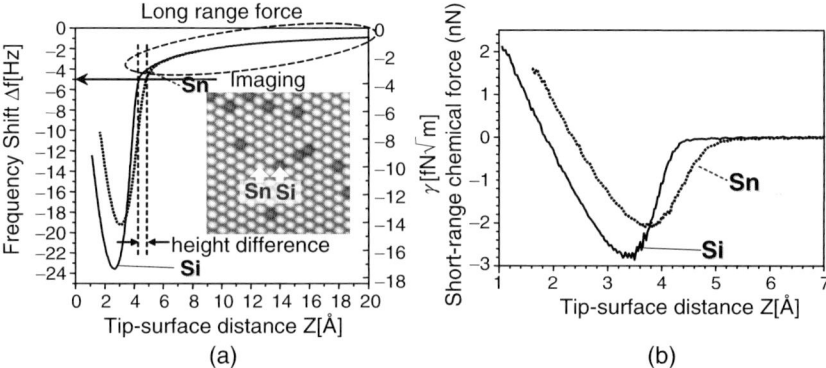

Fig. 10.4. (a) Frequency shift curves of the selected Si and Sn atoms indicated in the inset NC-AFM image, and (b) short-range chemical force curve obtained from (a) by using inversion procedure and then by subtracting long-range van der Waals force contribution [7]

of the selected Si and Sn adatoms [7]. As shown in Fig. 10.4b, Si adatom has maximum covalent bonding force stronger than Sn adatom in agreement with the common knowledge. Further, from obtained site-specific force curves of the selected Si and Sn adatoms shown in Fig. 10.4b, we can comprehend that Sn adatom has atom position higher than Si adatom, and/or covalent bonding radius larger than Si adatom. Therefore, for chemical identification, site-specific force spectroscopy is better method with abundant information than atom selective imaging.

Kelvin probe force microscopy (KPFM), which measures the contact potential difference (CPD) between the tip and sample surface, can also identify atom species such as intermixed Si and Sb adatoms in Si(111)$5\sqrt{3} \times 5\sqrt{3}$-Sb [8]. Besides, NC-AFM can detect topographic change due to the existence of subsurface B atom beneath Si(111) surface.

10.3 Roadmap

10.3.1 Recognition of Atom and Molecules; Inelastic Tunneling Spectroscopy

The road map of STM–IETS for the further development to be widely used chemical analysis instruments can be summarized as Fig. 10.5. Though this technique contains a large potential for many applications, there also are issues to be solved. One of the issues is that the detected vibrational modes are limited for the ones with high yield like as C–H stretching modes. It is necessary to detect all the vibrational modes for this technique to be chemical analysis tool. The key might be the improvement of the detection limit of inelastic component. In order to achieve that, further stability of the instruments are required including measurements at the temperature lower than 4 K. In addition there will be an increased interest for the detection vibration features of large molecules like biomolecule. The development of the techniques to measure such three-dimensional target will be necessary.

10.3.2 Future Prospect of Chemical Identification of Atoms by AFM

As shown in the roadmap of Fig. 10.6, the target of "Chemical Identification of Atom Species by AFM" developed from easier system with periodic atoms to more difficult one with intermixed aperiodic atoms. Until 2010, chemical

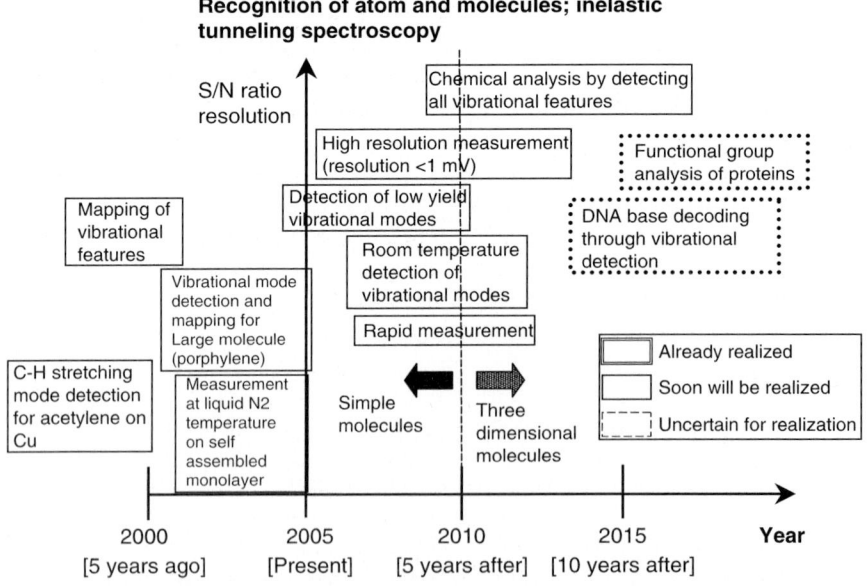

Fig. 10.5. Future development of inelastic tunneling spectroscopy for the single molecule chemical analysis

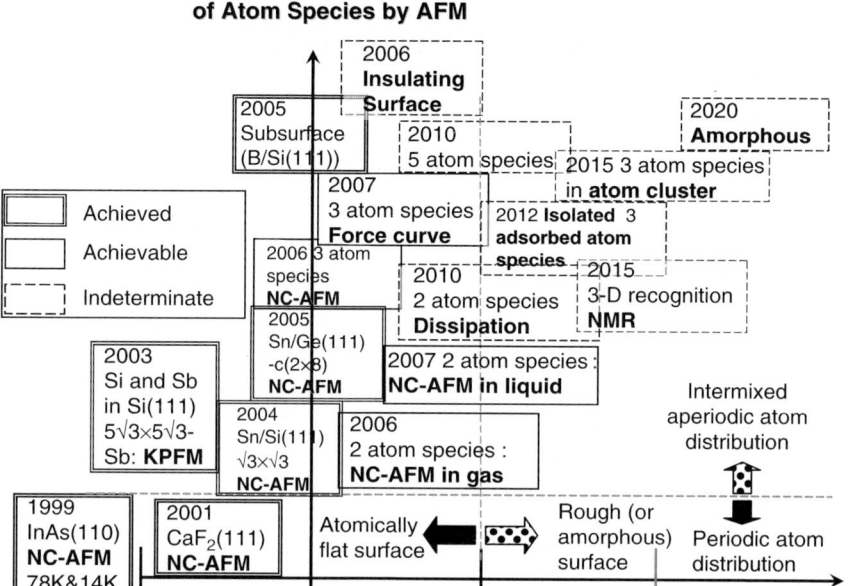

Fig. 10.6. Roadmap of chemical identification of atom species by AFM

identification of atom species in different environments such as gas and liquid or using different methods such as energy dissipation and force curve will become main subjects. Chemical identification of insulator and metal atoms will also become important subjects. After 2010, the target of chemical identification will move from easier sample with atomically flat surface to more difficult rough surface such as isolated adsorbed atom and atom cluster. Chemical identification of amorphous surface with not only intermixed aperiodic atoms but also various coordinations will also become important subjects. Besides, universal chemical identification of individual atoms using magnetic resonance force spectroscopy will become important subjects.

References

1. B.N.J. Persson, A. Baratoff, Phys. Rev. Lett. **59**, 339 (1987)
2. T. Komeda, Prog. Surf. Sci. **78**, 41 (2005)
3. Y. Sainoo et al., J. Chem. Phys. **120**, 7249 (2004)
4. J.A. Stroscio et al., J. Vac. Sci. Technol. A **6**, 499 (1988)
5. Y. Sugimoto et al., Appl. Surf. Sci. **241**, 23 (2005)
6. M. Abe, Y. Sugimoto, S. Morita, Nanotechnology **16**, S68 (2005)
7. Y. Sugimoto et al., Phys. Rev. B **73**, 205329 (2006)
8. K. Okamoto et al., Appl. Surf. Sci. **210**, 128 (2003)

Manipulation of Atoms and Molecules

Tadahiro Komeda, Seizo Morita, Shukichi Tanaka and Hirofumi Yamada

11.1 Manipulation of Atoms and Molecules: With the Use of STM Through Vibrational Excitation of Molecules

In the SPM roadmap 2,000 experiments of "quantum corral" done by Eigler's group [1], and H removal from Si substrate were already introduced [2]. These techniques are so excellent that it gave an impression that this area has been matured and no new physics would come out of this field. However, the idea turned out to be wrong and continuous and more exciting results have been produced in the last five years. Special emphasis should be made in the increase of the number of reports for the atomic scale manipulation and chemical reactions with vibrational excitation of adsorbates.

This shows an interesting contrast with the techniques employed in the last volume in which the majority of the manipulation were made through the van der Waals forces and strong electric field between the tip and the substrate. On the other hand, it has been theoretically predicted that the ladder climbing of vibrational mode is responsible for the Xe desorption which is one of the pioneering works of atom manipulation [3,4]. Chemical reactions caused by the vibrational ladder climbing has been extensively studied by using fast lasers in recent years revealing fascinating underlying chemistry and its application for coherent chemical reactions. Similar development of studies are expected with the STM manipulation. Recent progress in this field are reviewed.

First the mechanism of the tunneling current injection and the chemical reaction through the excitation of vibrational modes are briefly discussed. The excitation of a vibrational mode of a molecule in the STM junction is often explained using the resonant tunneling model in which an electron first tunnels from the tip to a molecule's resonant state then dissipated into the substrate. During the time period in which tunneling electrons stays at the resonant state, the molecule becomes an negative ion and the distance of nucleus tends to be expanded if compared to the neutral state. The motion can excite a vibrational mode of the molecule. Next it is convenient to use the idea of potential energy surface (PES) to understand how surface phenomena are induced by vibrational excitation. Let us consider a case where an molecule X desorbs from the surface A, for an example. The reaction coordinate should be the distance between X and A which is expressed as d here. The PES has a minimum at d_0 and forms a potential well around d_0. There appears a quantized vibrational modes in the well whose motion is along the reaction coordinates. In the desorption process, the mode should be the stretching mode between A and X. If the vibrational mode is excited to the quantum number whose energy is above the activation barrier then the desorption occurs. The phenomena should not be limited to the desorption; other phenomena such as hopping on the surface and chemical reaction of adsorbates should be induced with the same mechanism.

An example of the chemical reaction of a single molecule can be seen in a reaction from the *trans*-2-butene molecule to the butadiene molecule which is shown in Fig. 11.1 [5]. In the image of the backside panel of Fig. 11.1, we can see molecules of *trans*-2-butene (T) and butadiene (B) co-adsorbed on Pd(110) surface. When tunneling electrons are injected into the *trans*-2-butene molecule marked by an arrow, it can be changed into the molecule marked as P in the front panel, and it can be recognized that a butadiene molecule is formed by this process. The statistical investigation revealed that this process is induced by the excitation of the C–H stretching mode. The reaction order is two which indicates that the re-excitation of the $n = 1$ state to $n = 2$ state occurs before the quenching of $n = 1$ state. The $n = 2$ state is

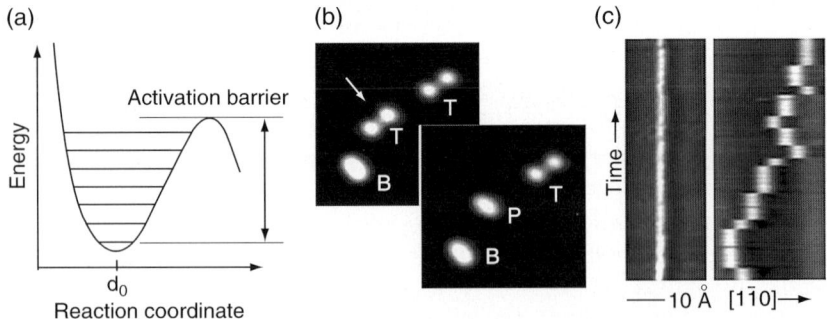

Fig. 11.1. (**a**) Schematic drawing of the potential energy surface using the reaction coordinate. (**b**) Chemical reaction of a single molecule of *trans*-2-butene (marked as T) into butadiene (B) molecule by the injection of tunneling electrons. Reaction product (P) is assigned as butadiene molecule. (**c**) Hopping of CO molecule by the injection of tunneling electron. The hopping probability shows a clear bias energy dependence. No hopping is induced in the left panel injecting electrons with the energy of 200 meV, while stochastic hopping is visible in case 300 meV electrons are injected shown in the right panel

located above the activation barrier, and C–H bond is broken by this ladder climbing followed by the conversion into a butadiene molecule.

We can also see a case where the energy is first stored in a high frequency mode of vibration which is distributed into multiple quanta of low-frequency mode of vibration which is directly related to the reaction coordinate and surface phenomena. The redistribution of the energy is caused by the anharmonic coupling of the two vibrational modes.

One can see such an example in the experiment of CO molecule manipulation on Pd(110) surface [6]. In the experiment, a CO molecule can change their bonding site like as hopping around the sites when tunneling electrons with the energy higher than the C–O stretching mode are injected into the molecule. In this process it is considered that high-frequency mode (C–O stretching mode) is first excited by tunneling electron though inelastic process. The vibrational mode is in the major part quenched by exciting the electron–hole pair of the substrate metal. However, a certain portion of the excited C–O stretching vibrations can be damped by exciting multiple low-frequency mode vibrational modes to vibrational state whose energy is higher than the activation barrier for the hopping. In the process, the anharmonic coupling between the high-frequency mode and the low-frequency mode is responsible.

Recently control of the path of the damping of the high frequency mode is studied on the system on NH_3 on Cu(111) surface [7]. In the desorption of the molecule, the hopping of the molecule is competing and by selecting the high frequency mode one can choose the desired phenomena.

11.2 Manipulation of Atoms and Molecules by AFM

11.2.1 Atom Manipulation by AFM

In 2000 when the first roadmap, "SPM Roadmap 2000" in Japanese, was published, atom manipulation by AFM was not achieved yet. At that time, "SPM Roadmap 2000" predicted that "Fundamental experiment for atom manipulation with AFM will proceed in 2003," and that "Manipulation of atoms and molecules with AFM will proceed in 2010." As introduced in Chap. 1.1, however, these future prospects were accelerated. To be specific, "Fundamental experiment of atom manipulation with AFM," was achieved in 2002, one-year earlier than the future prospect in "SPM Roadmap 2000," where Si adatom in Si(111)7×7 at low temperature of 9.3 K was mechanically extracted by the vertical atom manipulation using the mechanical vertical contact between the tip apex Si atom and the surface Si adatom [8]. Besides, another one of the future prospect on AFM, "Manipulation of atoms and molecules with AFM," was achieved in 2003 with full reproducibility, seven years earlier than the future prospect in "SPM Roadmap 2000," where Si adatom in Si(111)7×7 at low temperature of about 80 K was mechanically removed and then repaired (deposition of Si atom from the tip apex to the created

Si adatom vacancy) by the vertical atom manipulation using the mechanical vertical contact between the tip apex Si atom and the selected surface site [9]. Further, in 2005, single atom adsorbed on Ge(111)-c(2×8) substrate was laterally manipulated one by one according to the surface registry along [1,−1,0] crystal axis using the raster scan method [10].

Moreover, in the same year 2005, lateral atom-interchange manipulation phenomenon, which interchanges the embedded heterogeneous atom such as the selected Sn adatom embedded in Ge(111)-c(2×8) substrate with the selected adjacent Ge adatom, was discovered at room temperature (RT) [11]. Such tip-induced lateral atom-interchange manipulation was understood as the tip-induced directional thermal diffusion of the selected atom. Using such newly discovered lateral atom manipulation method that can be applicable even to the system consisted of multiatom species, "Atom Inlay" [11], that is the embedded atom letters "Sn" (the symbol of tin atom) consisted of 19 Sn atoms embedded in Ge(111)-c(2×8) substrate, was successfully created as

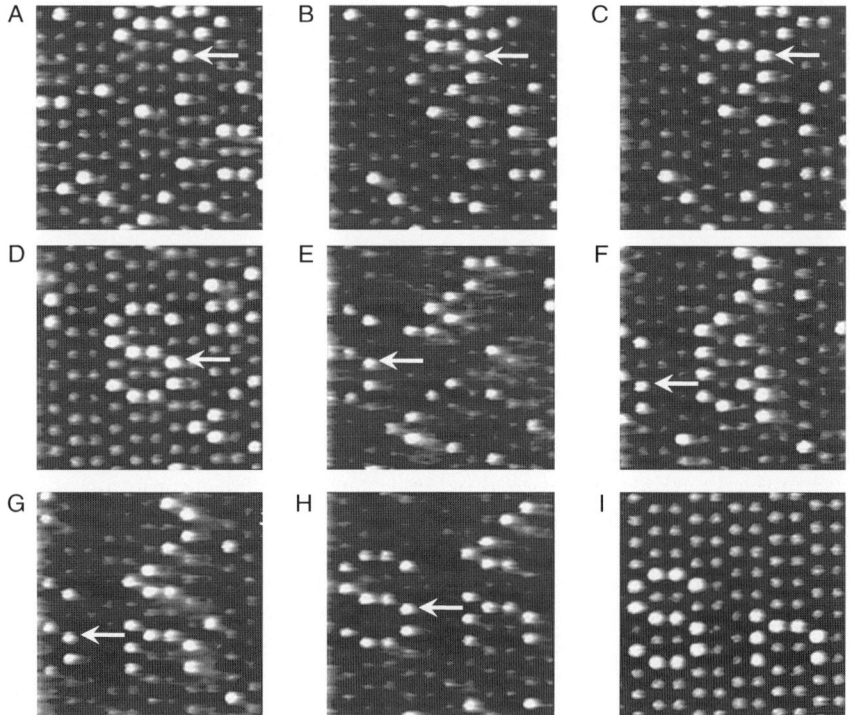

Fig. 11.2. Selected successive NC-AFM images during successive fabrication process (A=>B=>C=>///=>H=>I) of "Atom Inlay," i.e., embedded atom letters "Sn," which needed 9 h during 120 times of atom-interchange lateral manipulation of Sn and Ge atoms [11]

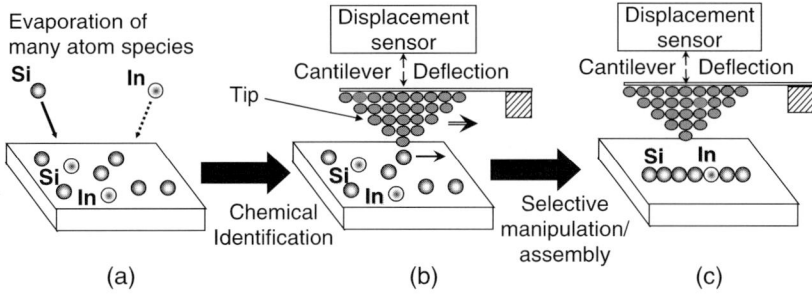

Fig. 11.3. Artificial nanostructuring by (**a**) evaporation of many atom species on the substrate, (**b**) chemical identification by AFM, and then (**c**) selective atom manipulation and assembly by AFM

shown in Fig. 11.2 at RT. By this achievement, practical bottom up process as shown in Fig. 11.3, that is chemical identification of atom species, following manipulation of selected atoms and then assembly of artificial nanostructure, with novel functions, consisted of many atom species, became accessible, for the first time, at room temperature.

Besides, another one of the future prospect on AFM, "True atomic resolution under liquid environment," was achieved in 2005, five-years earlier than the future prospect in "SPM Roadmap 2000" [12]. Such rapid progresses of AFM will at least partly owe to the stimulation by the future prospect in "SPM Roadmap 2000."

11.2.2 AFM Manipulation of Organic Molecules

In the early development of techniques, most AFM experiments of organic molecule systems[1] were limited to observations of their conformational features and physical properties of the sample surface. The achievements to use this technique as a manipulator of individual molecules have appeared in this decade with their recent technical progresses. Although various methodologies to configure the local properties and shapes of materials, such as local scratching, local indentation, anode oxidation, have been attempted, the use of AFM probe techniques to manipulate nanometer scale individual molecules has been the recent focus of innovative nanotechnology research.

Advantages of using AFM probe technique as a precise manipulator for molecules are following:

1. The effective size of an AFM probe is smaller than that of individual molecule units, and the probe can directly access inner parts of molecules.

---

[1] In this section, "molecules" means "organic molecules with their unit weights of 100–1,000."

2. Possibility for manipulation and modification of target molecules via mechanical interaction between the AFM probe and molecules.
3. The results of probe manipulation can be checked in situ using the same AFM scanning system.
4. The electric current flow between the probe and molecules can be used as a parameter of manipulation independently to the AFM feedback motion.
5. More complicated manipulation is possible by using the chemical interaction between molecules and a chemically modified AFM probe.

Especially, it is a versatile feature for nanometer scale manipulation that (1) and (3) is possible. Considering recent achievements and expected progress in this research field, key topics and trends are summarized in Fig. 11.4.

Each topic is roughly classified along two technical directions, mechanical or chemical treatments, which are important schemes in understanding the mechanisms of AFM manipulation of molecules. At present, there are only a few successful examples of AFM manipulation of molecules that take full advantage of the benefits of AFM manipulation function, especially its high spatial resolution.

Dip pen nanolithography (DPN) technique is one of the promising techniques for precise positioning of target molecules on the surface [13]. With DPN technique, prior to the lithographic process by an AFM probe, the apex

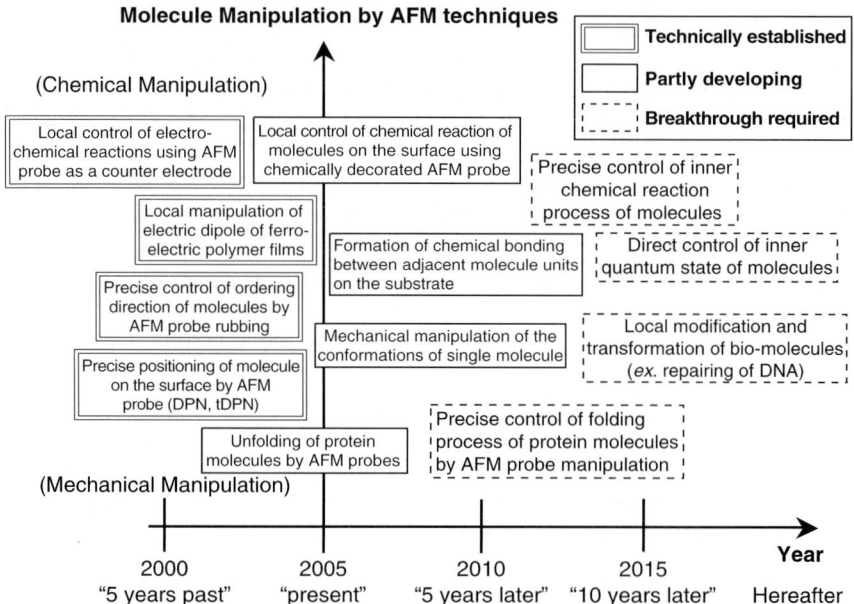

Fig. 11.4. Roadmap concerning the AFM manipulation of organic molecules

of the AFM probe is dipped into specially designed molecule ink, in which a number of the target molecules are dissolved. This coated AFM probe is then approached to the surface and draws a line pattern on the surface using an AFM scanning mechanism. Although the mechanism itself is very simple, this technique enables very small patterns of molecules with widths of 100–200 nm to be drawn very precisely and flexibly on the substrate. Recently, the precision of the drawn pattern and its application has been drastically improved as a result of combining specially engineered AFM probe with small heating units and solid ink that melts quickly by heating thermal dip pen nanolithography (tDPN) [14]. By choosing an appropriate solvent and heating mechanism, a user will be able to use this drawing system in UHV condition, and that very precise drawing by only a few molecules might be possible in the near future.

The capabilities of AFM manipulation techniques have raised considerable interest for the precise conformational control of carbon nanotubes (CNT), whose unique properties are fascinating and promising for various applications [15]. Most experiments are currently performed using a classical system based on contact-mode AFM, and that the functions and usability of a nanoscale manipulator is not optimized. There has been some recent progress in improving the usability of AFM manipulation by using a haptic system [16]. This system is an interactive human interface device that connects a human's sense to manipulation motion, and this method is expected to be one of the important technical elements for AFM manipulation techniques. There are many possibilities for technical improvements in these manipulation techniques, for example, by adopting dynamic mode AFM feedback operation and applying local external fields, like electric voltage. Another important example of the significance of this method is its application for manipulating single biomolecules. In 2001, Müller et al. succeeded in unfolding a polypeptide chain stacked as a three-dimensional composite in HIP layer by catching and pulling the peptide chain with an AFM probe [17]. This unfolding process of protein molecules can be visualized by analyzing the AFM force curve recorded during the unfolding. The effect of AFM manipulation on HIP layers can be easily checked by in situ AFM observation, which confirms experimental details. Such experiments are almost impossible without AFM-based techniques, and the technique is receiving increased attention as an important tool for analyzing the folding process and mechanisms of three-dimensional proteins.

There are also a number of novel trials to improve the macroscopic properties of thin films of organic molecules by nanoscale molecular manipulation using AFM-based techniques. Among them, nanorubbing techniques performed by AFM probe scanning has potential industrial applications. In this technique, the geometrical ordering of individual molecules on the surface is controlled and aligned in specific direction using precise AFM probe scanning. There are some successful reports for ferroelectric molecular thin films and liquid crystals [18, 19].

11.3 Roadmap

11.3.1 Manipulation of Atoms and Molecules: with the Use of STM Electrons

The energy of vibrational mode of molecules accidentally coincide with one of the tunneling electrons used in STM experiment. State-to-state chemical reaction investigation is developed with the improvement of the laser. The manipulation of the adsorbate by the use of the tunneling electrons will be developed following the success using the laser light. So far there is no control of the phase of the tunneling current; but as the coherent control of chemical reaction is widely examined in laser chemistry, one should soon start control of the phase of the injected electrons (see Fig. 11.5).

11.3.2 Future Prospect of Atom Manipulation by AFM

As a result of the achievement of "Atom Inlay," that is, embedded atom letter in Fig. 11.2(I), the target of "mechanical atom manipulation by AFM" in UHV developed from simpler manipulation of homogeneous atom to more complex manipulation of many atom species, which needs chemical identification of atom species, as shown in the roadmap of Fig. 11.6. Until 2010,

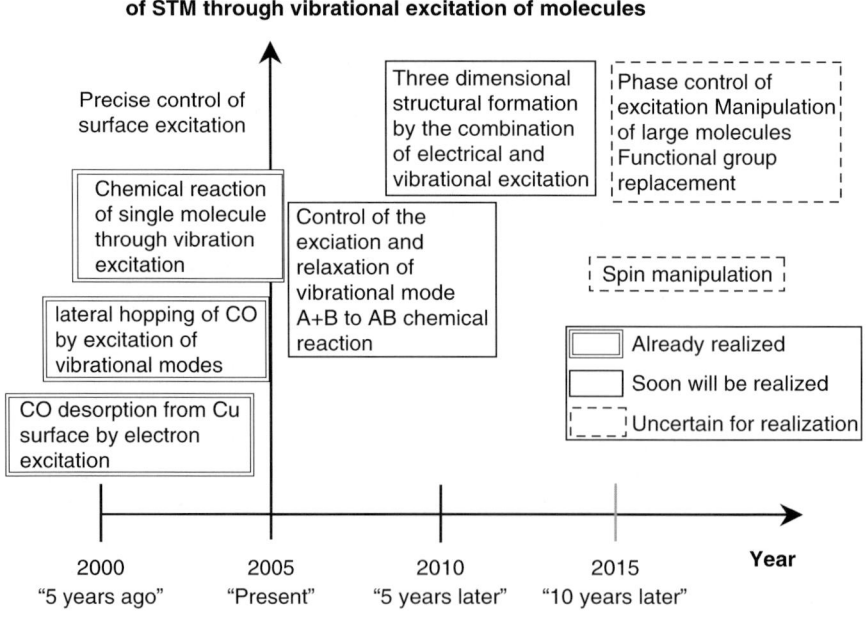

Fig. 11.5. Future of the manipulation of atoms and molecules through vibrational excitation of molecules

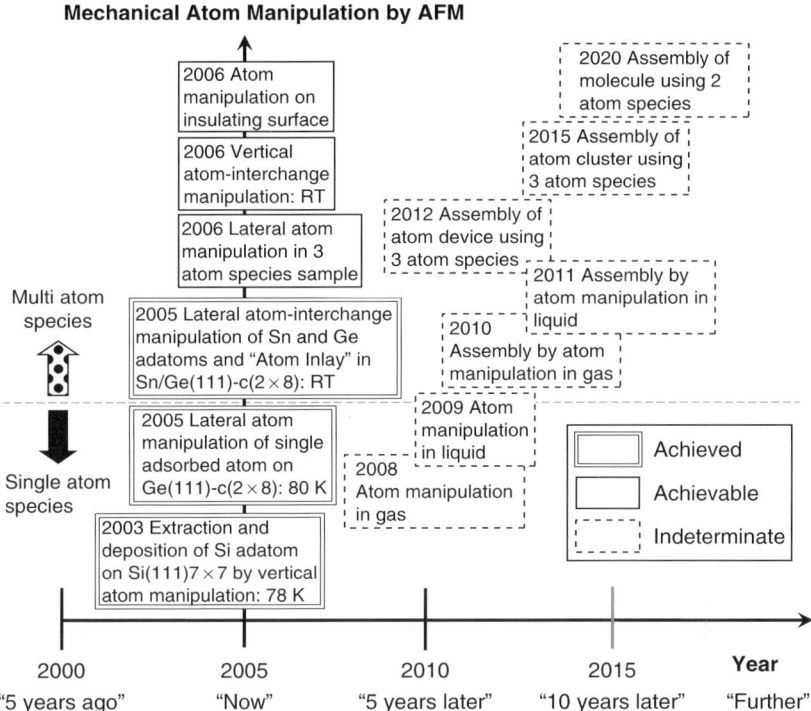

Fig. 11.6. Roadmap of mechanical atom manipulation by AFM

chemical identification of intermixed heterogeneous insulator and metal atoms and following selected atom manipulation will become the main subjects in UHV. Besides, manipulation of single atom in gas and liquid will be important subjects. After 2010, aiming search and creation of complex nanostructures with novel functions consisted of more than two atom species, assembly of atom cluster and atom devices at room temperature consisted of many atom species will become the main subjects. Selected atom manipulation of inter-mixed many atom species and following artificial nanostructuring in gas and liquid will also become important subjects. Further, development of technique that assembles arbitrary molecules using observation, chemical identification, and manipulation of atoms and molecules is the most challenging subject of atom manipulation.

11.3.3 Future Prospect of AFM Manipulation of Organic Molecules

Organic molecules are one of the fundamental elements in establishing nanomaterial technology. The most promising method to treat such small, less conductive and fragile materials is, undoubtedly, AFM related techniques.

Considering the current situation in this research field, AFM related nanoprobe techniques will drastically improve their application in actual situations and would improve their functions from "observations" to "constructions" in the next five years. New trends in AFM technologies, such as precise probe positioning system by using closed-loop techniques, numerically controlled feedback system using DSP electronics [20], and so on, should contain important technical elements that enable the development of practical nanoscale precision manipulators for molecules. For this purpose, a good understanding is required of the details of the chemical and physical interactions between AFM probes and the molecules on the surface to control their features. To develop novel and innovative AFM technique as fundamental tools for nanomaterial designing, we have to customize the various properties inside of the molecules with an AFM probe. However, it will take longer than five years to establish enough technical expertise to achieve nanomaterial technology.

References

1. M.F. Crommie, C.P. Lutz, D.M. Eigler, Science **262**, 218 (1993)
2. T.C. Shen et al., Science **268**, 1590 (1995)
3. D.M. Eigler, C.P. Lutz, W.E. Rudge, Nature **352**, 600 (1991)
4. R.E. Walkup, D.M. Newns, P. Avouris, Phys. Rev. B **48**, 1858 (1993)
5. Y. Kim, T. Komeda, M. Kawai, Phys. Rev. Lett. **89**, 126104 (2002)
6. T. Komeda et al., Science **295**, 2055 (2002)
7. J.I. Pascual et al., Nature **423**, 525 (2003)
8. S. Morita, R. Wiesendanger, E. Meyer (eds.), *Noncontact Atomic Force Microscopy* (Springer, Berlin Heidelberg New York, 2002) 3.10
9. N. Oyabu et al., Phys. Rev. Lett. **90**, 176102 (2003)
10. N. Oyabu et al., Nanotechnology **16**, S112 (2005)
11. Y. Sugimoto et al., Nat. Mater. **4**, 156 (2005)
12. T. Fukuma et al., Rev. Sci. Instrum. **76**, 053704 (2005)
13. K.-B. Lee et al., Science **295**, 1702 (2002)
14. P.E. Sheehan et al., Appl. Phys. Lett. **85**, 1589 (2004)
15. T. Hertel, R. Martel, P. Avouris, J. Phys. Chem. B **102**, 910 (1998)
16. M.R. Falvo et al., Microsci. Microanal. **4**, 504 (1999)
17. D.J. Müller, W. Baumeister, A. Engel, Proc. Natl. Acad. Sci. USA **96**, 13170–13174 (1999)
18. K. Kimura et al., Appl. Phys. Lett. **82**, 4050 (2003)
19. J.-H. Kim, M. Yoneya, H. Yokoyama, Nature **420**, 159 (2002)
20. Abbreviation of Digital Signal Processor. More information can be found in: S.W. Smith, The Scientist and Engineer's Guide to Digital Signal Processing (California Technical Publishing, California 1997)

Multiprobe SPM

Shuji Hasegawa

12.1 Present Status

The multiprobe scanning probe microscope (SPM), in which several tips or cantilevers are independently driven and arranged in arbitrary configurations on samples surfaces, has recently attracted considerable attention as a very versatile tool for electrical characterization at nanometer scales. The SPM tips/probes are employed as current sources, voltage pick-up probes, and field-gate electrodes as well as tweezers for structure manipulations. Several groups are developing different types of multiprobe SPMs [1–7], and some companies begin to deliver the products [8]. These commercial machines are mainly for testing electrical characteristics of nanometer-scale electronic devices, and regarded as a tool evolved from conventional electrical probers. Some of them are for electrical measurements of biological cells and proteins. In order to control the contact pressure between the probes and sample surfaces, many of them have ability of atomic force microscopy not only for imaging, but also for electrical measurements with conductive cantilevers. But they do not necessarily have atomic resolutions, and control precision of tip/probe positions is poorer than 10 nm. The apparatus and operation system are not yet fully developed, and still have much room for evolution in many aspects. Especially the operation system for controlling the multiprobes with atomic precisions as an organic whole is still lacking, and therefore it seems that the true value of multiprobe SPM is not yet realized.

Figure 12.1a shows a schematic of the four-tip scanning tunneling microscope (STM) apparatus in an ultrahigh vacuum (UHV) chamber developed at the University of Tokyo, and Fig. 12.1b shows a photograph of the goniometer stage on which the sample and four sets of scanners are mounted [9, 10]. The four tips of STM are driven independently under scanning electron microscope (SEM) for positioning the tips precisely with arbitrary arrangements on specified areas on the sample surface. Each tip points to the sample at the center with 45° from the sample surface, and is driven by a special type of a piezo-scanner for fine positioning and by three sets of piezo-actuators

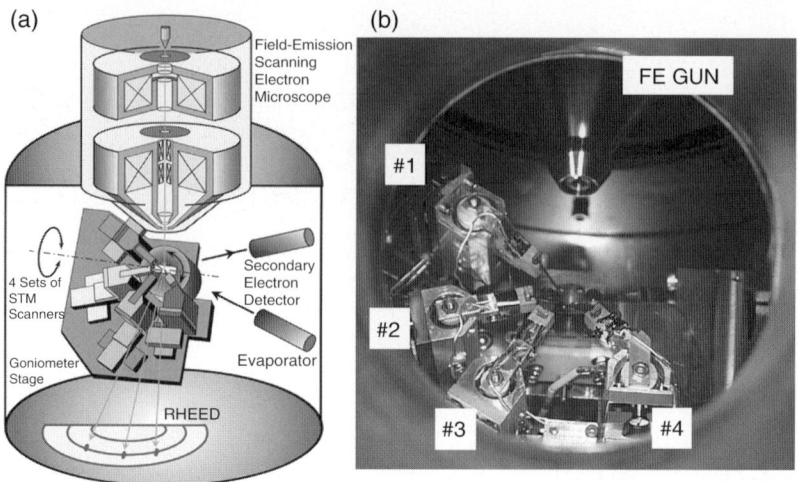

Fig. 12.1. (a) Schematics of the independently driven four-tip STM, installed in an UHV-SEM-RHEED system. (b) A photograph of the goniometer stage on which a sample and the four-tip scanners are mounted under the SEM column [9, 10]

(Microslide, Omicron) for coarse motion. The goniometer stage enables parallel shifts in three directions and tilt rotation with respect to the SEM electron beam. The sample can be rotated azimuthally by 360° with respect to the stage. These positioning mechanisms enable fine adjustments with respect to the SEM electron beam, required to perform reflection-high-energy electron diffraction (RHEED) and scanning reflection electron microscopy (SREM) observations of the sample surface simultaneously. These supplementary electron microscopy/diffraction techniques are indispensable not only for positioning the four tips properly, but also for confirming the surface structures of sample. The STM tips and sample can be exchanged and installed by transfer rods from load-lock chambers without breaking vacuum.

This apparatus enables usual STM operation by each tip independently, and also four-point probe (4PP) conductivity measurements with various probe arrangements and spacing. The four tips approach the sample surface simultaneously with feedback control by tunnel-current detection. After that, the tips are brought into direct contacts with the sample surface, and then the 4PP conductivity measurement is performed. The preamplifier is switched from the tunnel-current mode to the 4PP conductivity measurement mode. Control system for the four-tip STM is still in its infancy. Each tip is independently controlled, but not in an integrated way. If the positions of all tips are controlled by their xyz coordinates at nanometer-scale precision by a single controller, we do not need SEM for tip positioning anymore. A method for navigating two STM tips is developed by using a special type of sample [11].

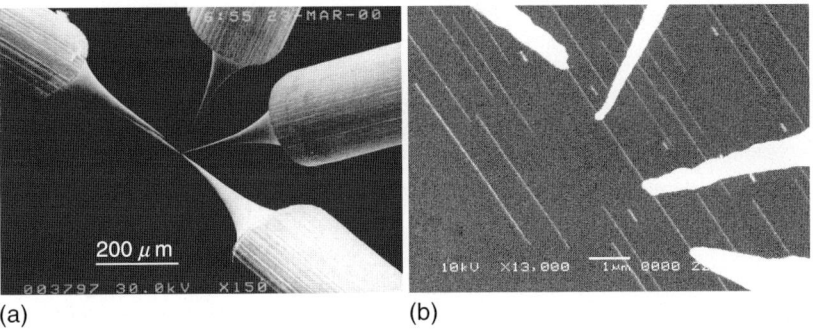

Fig. 12.2. (a) SEM images of the four W tips in the four-tip STM [9]. (b) The four tips contacting a Co-silicide nanowire on an Si(110) surface [12]

Figure 12.2a shows an SEM image of four tungsten tips in the four-tip STM [9, 12]. The probe spacing can be changed from ca. 100 nm to 1 mm, and arranged in arbitrary ways such as in linear or in a square with equidistant probe spacings [13]. These four tips are used for the 4PP conductivity measurements of microscopic regions and objects. When the probe spacing is reduced on the order of microns, we can measure the electrical conductivity through the topmost atomic layers on a crystal with high sensitivity [14, 15] as well as individual microscopic objects such as nanowires [12]. When the four tips are arranged in a square on a sample surface, we can measure the anisotropy of conductivity [13].

12.1.1 Improvements

The following two issues should be improved from technical points of view in order to make the multiprobe SPM a more versatile tool for nanometer-scale measurements.

1. *Control system.* Many of the multiprobe SPMs need an auxiliary microscope such as SEM and optical microscope to observe and position the tips/probes in the designed arrangements on a sample surface. And in many cases the tips/probes are independently driven with separate sets of controllers without mutual communication. These make the operation very troublesome. Therefore, a user-friendly controller, by which the multiprobes are controlled with nanometer precisions in integrated ways by a single computer, is highly desired. Furthermore, it becomes much more convenient if we can control each tip by its xyz coordinates. For this purpose, we need a so-called closed-loop system in which the probe movements driven by scanners are simultaneously measured by some kinds of displacement sensors, and the results are used for feedback of tip positioning. Such a controller is recently produced experimentally, while the tip-positioning precision is not yet enough for the nanometer-scale measurements.

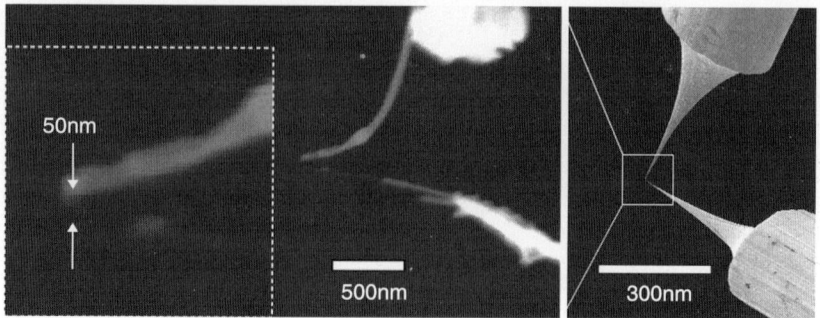

Fig. 12.3. SEM images of the two metal-coated carbon nanotubes arranged with ca. 50 nm spacing in the four-tip STM [17]

2. *Tips/probes.* The minimum spacing between two tips is determined by the radius r of the tip end; it is impossible to bring the two tips close to each other less than $2r$, because the two tips touch each other. In the case of electrochemically etched tungsten tips which are usually used for STM, $r \sim 50$ nm, which means that the minimum tip spacing is ca. 100 nm. Therefore, it is necessary to utilize much thinner tips such as carbon nanotubes (CNTs) and whiskers. Figure 12.3 shows two CNT tips in the four-tip STM, by which we can make the two tips approach each other less than 50 nm [16–18]. A multiwalled CNT is glued on the end of a W tip, and wholly coated by a thin W layer to make the junction between the CNT and W supporting tip conductive. Such coating by a thin metal film is indispensable to make the tip conductive enough for the STM and electrical measurements. With this technique by utilizing CNTs, we will be able to reach the minimum tip distance around 15 nm. In addition to metal layers, it is possible to coat the CNT tips with other materials such as dielectric, magnetic, and superconducting materials [19, 20], the multiprobe SPM will have various uses in different ways not only for electrical conductivity measurements.

12.1.2 Roadmap

If the technical issues mentioned earlier are improved and the probe spacing reaches down to ca. 10 nm routinely, various measurements and applications by the multiprobe SPM will become possible as shown in Fig. 12.4. We will be able to measure the electrical conductance of individual nanometer-scale objects such as DNA molecules, atomic chains, and nanodots. In the measurements, then, the influence of tip contact will be a serious problem. The electrical probes with direct contact to the sample will easily disturb the states and structures of such nano-objects. To avoid this disturbance, the simultaneous tunneling contact of multiprobes will be indispensable.

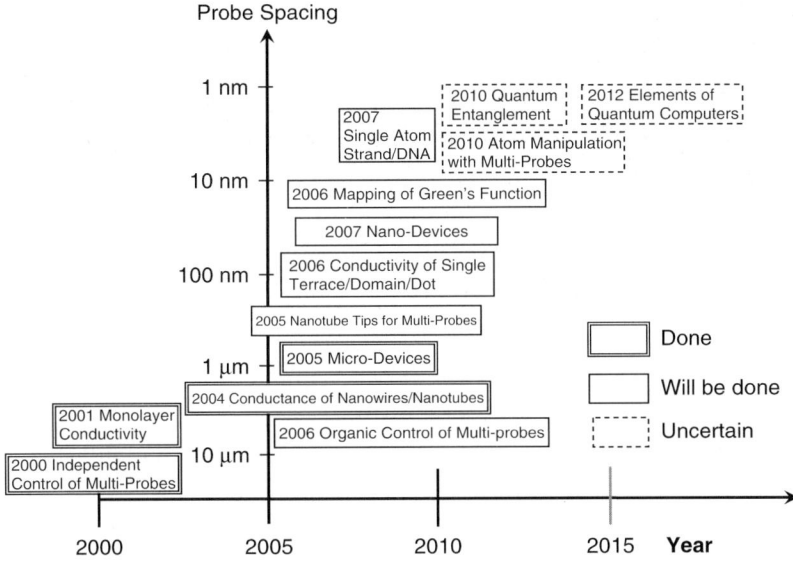

Fig. 12.4. Future prospect about the multiprobe SPM

When the tip spacing is comparable to the coherence length of carriers in the sample, a new type of measurement, i.e., real-space mapping of Green's function, will be possible by the multitip STM [21,22]. This "Green's function STM" measures the change of tunneling current through a tip during changing the tunneling bias voltages of other tips. Such nonlocal phenomena including electron correlation effects can be measured when the tips are brought close to each other in a range of carrier coherence length. With this new type of multiprobe SPM, it will be possible to measure quantum entangled states in nanostructures which will be useful as elements of quantum computers.

The multiprobe SPM will be also utilized for structural modifications and atomic/molecular manipulations. With use of multiprobes, such structure constructions and characterizations will be simultaneously possible.

References

1. S. Tsukamoto, B. Siu, N. Nakagiri, Rev. Sci. Instrum. **62**, 1767 (1991)
2. M. Aono, C.-S. Jiang, T. Nakayama, T. Okuda, S. Qiao, M. Sakurai, C. Thirstrup, Z.-H. Wu, J. Surf. Sci. Soc. Jpn. **19**, 698 (1998), in Japanese
3. H. Okamoto, D.M. Chen, Rev. Sci. Instr. **72**, 4398 (2001)
4. H. Grube, B.C. Harrison, J. Jia, J.J. Boland, Rev. Sci. Instr. **72**, 4388 (2001); Rev. Sci. Instr. **73**, 1343 (2002) [ERRATA]
5. H. Watanabe, C. Manabe, T. Shigematsu, M. Shimizu: Appl. Phys. Lett. **78**, 2928 (2001); **79**, 2462 (2001)

6. K. Takami, M. Akai-Kasaya, A. Saito, M. Aono, Y. Kuwahara, Jpn. J. Appl. Phys. **44**, L120 (2005)
7. M. Ishikawa, M. Yoshimura, K. Ueda, Jpn. J. Appl. Phys. **44**, 1502 (2005)
8. Omicron NanoTechnology GmbH (http://www.omicron.de/), MultiProbe, Inc (http://www.multiprobe.com/), Zyvex Co. (http://www.zyvex.com/)
9. I. Shiraki, F. Tanabe, R. Hobara, T. Nagao, S. Hasegawa, Surf. Sci. **493**, 643 (2001)
10. S. Hasegawa, I. Shiraki, F. Tanabe, R. Hobara, Current Appl. Phys. **2**, 465 (2002)
11. H. Okamoto, D.M. Chen, J. Vac. Sci. Technol. A **19**, 1822 (2001)
12. H. Okino, I. Matsuda, R. Hobara, Y. Hosomura, S. Hasegawa, P.A. Bennett, Appl. Phys. Lett. **86**, 233108 (2005)
13. T. Kanagawa, R. Hobara, I. Matsuda, T. Tanikawa, A. Natori, S. Hasegawa, Phys. Rev. Lett. **91**, 036805 (2003)
14. S. Hasegawa, I. Shiraki, F. Tanabe, R. Hobara, T. Kanagawa, T. Tanikawa, I. Matsuda, C.L. Petersen, T.M. Hansen, P. Boggild, F. Grey, Surf. Rev. Lett. **10**, 963 (2003)
15. S. Hasegawa, I. Shiraki, T. Tanikawa, C.L. Petersen, T.M. Hansen, P. Goggild, F. Grey, J. Phys.: Cond. Matters **14**, 8379 (2002)
16. T. Ikuno, M. Katayama, M. Kishida, K. Kamada, Y. Murata, T. Yasuda, S. Honda, J. Lee, H. Mori, K. Oura, Jpn. J. Appl. Phys. **43**, L644 (2004)
17. Y. Murata, S. Yoshimoto, M. Kishida, D. Maeda, T. Yasuda, T. Ikuno, S. Honda, H. Okada, R. Hobara, I. Matsuda, S. Hasegawa, K. Oura, M. Katayama, Jpn. J. Appl. Phys. **44**, 5336 (2005)
18. S. Yoshimoto, Y. Murata, R. Hobara, I. Matsuda, M. Kishida, H. Konishi, T. Ikuno, D. Maeda, T. Yasuda, S. Honda, H. Okada, K. Oura, M. Katayama, S. Hasegawa, Jpn. J. Appl. Phys. **44**, L1563 (2005)
19. M. Kishida, H. Konishi, Y. Murata, D. Maeda, T. Yasuda, T. Ikuno, S. Honda, M. Katayama, S. Yoshimoto, R. Hobara, I. Matsuda, S. Hasegawa, e-J. Surf. Sci. Nanotech. **3**, 417 (2005)
20. T. Ikuno, T. Yasuda, S. Honda, K. Oura, M. Katayama, J.-G. Lee, H. Mori, J. Appl. Phys. **98**, 114305 (2005)
21. Q. Niu, M.C. Chang, C.K. Shih, Phys. Rev. B **51**, 5502 (1995)
22. J.M. Byers M.E. Flatte: Phys. Rev. Lett. **74**, 306 (1995)

13

AFM Measurement in Liquid

Hirofumi Yamada

13.1 Demand for AFM Imaging in Liquid

Atomic force microscopy (AFM), working in vacuum, air, and even in liquid, does not have any restrictions on imaging environments. In addition, no special treatment for imaging samples such as staining or metal coating, which are common techniques in electron microscopy imaging, is required. Because of these remarkable advantages over other high-resolution imaging methods, AFM applications to atomic/molecular-scale *in vivo* imaging of biological samples in physiological, aqueous environments are greatly expected.

There were a large number of AFM studies on biological samples in liquid by contact mode AFM. However, one can often experience that the scanning process in contact mode modifies or damages soft biological samples, most of which are adsorbed onto a substrate by weak van der Waals interaction. Thus dynamic force microscopy (DFM), which can drastically reduce perturbative interactions, is now a most common method for imaging the biological samples [1]. There are two working modes in DFM. One is AM-AFM (often referred to as tapping mode AFM) using amplitude modulation (AM) detection, which usually works in the intermittent contact regime, and the other is FM-AFM using frequency modulation (FM) detection, which are mainly used in the noncontact regime. Average interaction/contact forces in typical imaging conditions of each mode are heavily reduced compared to the case of contact mode AFM and hence sample damages can be avoided.

On the other hand, DFM imaging in liquid has some difficulties such as unwanted vibration excitation of an AFM cantilever and a large decrease in force sensitivity due to a low Q-factor, as explained in Sect. 13.3. In this chapter the present status of AFM imaging in liquid and its technology roadmap are described.

13.2 Dynamic Mode AFM Imaging in Liquid

13.2.1 AM-AFM and Q-Control Method

A cantilever is vibrated at a fixed frequency near the resonance frequency in AM-AFM (tapping mode AFM). The amplitude of the cantilever vibration is varied depending on instantaneous contact interactions in each oscillation cycle of the cantilever. Surface topographic data is obtained by measuring the change in the amplitude, from which the term "AM-AFM" derives [2].

Dynamic mode AFM utilizing cantilever having a sharp resonance in air or vacuum was originally developed for increasing the force sensitivity as well as reducing the contact force [3]. In contrast, when the cantilever is oscillated in liquid, the Q-factor is decreased due to the viscosity and inertia resistance of liquid, which causes a large decrease in the force sensitivity and an unstable cantilever oscillation. In addition, it is pointed out that soft biological samples can be damaged in a low-Q-environment in the same amplitude condition, i.e., amplitude set point [4].

Q-control method, where the cantilever vibration is controlled by use of phase-shifted amplitude feedback, was developed for overcoming these difficulties [5]. Figure 13.1 shows a schematic diagram of the Q-control method. The phase of the cantilever oscillation signal is shifted by θ ($\sim \pi/2$) using a phase shifter or a differential circuit. The signal is amplified (Gain G) and then added to the driving signal ($F_0 e^{i\omega t}$) for the cantilever oscillation. Equation of motion for the cantilever mechanical displacement $z(t)$ is described in the complex form (taking it into account that the phase is controlled):

$$m\ddot{z} + m\gamma\dot{z} + kz = F_{\mathrm{ts}}(z) + F_0 e^{i\omega t} + Ge^{i\theta}z, \tag{13.1}$$

where m, ω_0, k ($= m\omega_0^2$), γ ($= \omega_0/Q$), and $F_{\mathrm{ts}}(z)$ are the effective mass in liquid, resonance angular frequency, spring constant, damping constant of the cantilever, and interaction force between the tip and the sample. When the

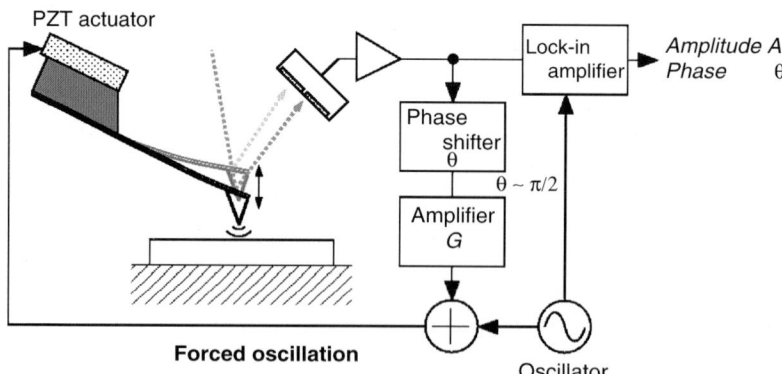

Fig. 13.1. Schematic of experimental setup for Q-control method

tip is located far from the sample surface ($F_{ts}(z) = 0$) and $\theta = \pi/2$, $z = Ae^{i\omega t}$ can be a solution. The z terms having an imaginary coefficient are related to the effective damping. In this case the effective damping constant γ_{eff} is expressed as

$$\gamma_{\text{eff}} = \gamma - \frac{G}{m\omega_0\omega}. \tag{13.2}$$

This effective Q-factor (Q_{eff}: $\omega_0/\gamma_{\text{eff}}$) can be changed to be any value by controlling G (increased when $G > 0$, decreased when $G < 0$).

The increase of the Q-factor is directly related to the average load to the sample $\langle F_{ts} \rangle$ [4]. Under the approximation $\langle F_{ts}z \rangle = -A\langle F_{ts} \rangle$, the magnitude of $\langle F_{ts} \rangle$ can be written by [6, 7]

$$\langle F_{ts} \rangle = \frac{k}{Q_{\text{eff}}}(A_0^2 - A_{\text{sp}}^2)^{1/2} \approx kA_0 \frac{1}{Q_{\text{eff}}}\left(2\frac{\delta A}{A_0}\right)^{1/2}, \tag{13.3}$$

where A_0 and A_{sp} are the amplitude of the free oscillation and the amplitude used in AFM imaging and $\delta A = A_0 - A_{\text{sp}}$. The terms with the second and higher order of $\delta A/A_0$ were ignored. The equation clearly indicates that the increase of Q_{eff} leads to the decrease in $\langle F_{ts} \rangle$.

When the cantilever vibration is excited in liquid, various spurious peaks usually appear in the vibration spectrum, which strongly depends on how the cantilever is vibrated. In this case even the identification of the cantilever resonance peak is sometimes difficult. Since the Q-control method enhances the resonance spectrum, the difficulty in the resonance identification process can be fairly reduced [8].

The signal level seems to be increased due to the apparent increase in Q-factor. However, since the artificial control of the Q-factor is also effective to the noise amplification with the same gain factor, the signal-to-noise ratio, which is determined by the thermal fluctuation, cannot be improved [9].

13.2.2 High-Resolution Imaging by FM-AFM

While the cantilever is vibrated at a fixed frequency in AM-AFM, the cantilever oscillation always tracks the resonance frequency in FM-AFM where the cantilever works as a mechanical resonator. Interaction forces between the tip and the sample cause a shift of the resonance frequency and hence the same shift of the cantilever oscillation frequency. This frequency shift is used as a feedback signal for the z tip/sample motion, which corresponds to the topographic information. There has been a great progress in FM-AFM. High-resolution AFM imaging of atomically flat samples in UHV environments is becoming a routine work. FM-AFM is now widely used as a high-resolution analysis method for various materials including semiconductors, insulators, and organic molecules.

Since FM-AFM requires an electromechanical resonator for its self-oscillation circuit, the high Q-factor of a cantilever is prerequisite for stable

AFM operation. It was not likely that FM-AFM worked in an environment where the Q-factor was very low such as in liquid. Recently the development of a narrow-band FM detector using a highly stable phase-locked loop circuit brought a great progress in FM-AFM imaging in liquid [10]. The measurement of a change in oscillation frequency by the FM detector is detected as a change of the oscillation phase. In this sense the FM detection is somewhat similar to the Q-control method, where the signal phase is controlled. In fact the average interaction force $\langle F_{ts} \rangle$ is also decreased by FM detection. The reduction ratio of $\langle F_{ts} \rangle$ approximates the ratio of a frequency shift to the resonance frequency, which can be 10^{-3}–10^{-4} for the ordinary imaging condition in FM-AFM.

Recently, true atomic resolution imaging in liquid have been successfully achieved by FM-AFM [11, 12]. The difficulty in imaging in liquid was overcome mainly by the use of the small amplitude mode ($A \ll 1\,\text{nm}$) and the noise reduction ($17\,\text{fm}\,\text{Hz}^{-1/2}$) in the cantilever deflection sensor. The force sensitivity is increased by FM detection with small amplitude oscillation because of the increase in the duration of the proximity interaction. Note that the small amplitude mode can be used only when the noise in the deflection sensor is sufficiently reduced down to a level of the thermal fluctuation of the cantilever. The phase noise in the FM detector is proportional to the ratio of the measurement noise to the oscillation amplitude.

A cleaved surface of a polydiacetylene ($poly$-PTS: 2,4-hexadine-1,6-diol bis[p-toluene sulfonate]) single crystal was imaged in pure water by FM-AFM (Fig. 13.2). The herring bone structures of the side groups (p-toluene sulfonate) was clearly detected. Figure 13.3 shows frequency shift vs. distance curves obtained on this sample. Each curve is clearly modulated with a period of about 0.2 nm in the positive frequency shift region, which is probably caused by a hydration shell structure.

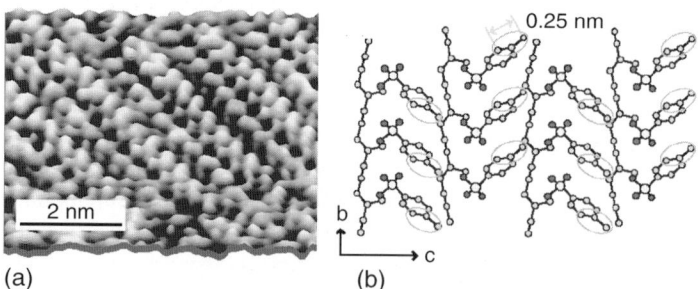

(a)   (b)

Fig. 13.2. (a) FM-AFM image of the bc plane of a polydiacetylene single crystal ($poly$-PTS) taken in water ($A = 0.20\,\text{nm}$, $\Delta f = +290\,\text{Hz}$, $Q = 27$, $f_0 = 140\,\text{kHz}$, $k = 42\,\text{N}\,\text{m}^{-1}$). (b) Crystal structure of the bc plane. In the bc plane, one side of the PTS side groups is located under the other side. Thus, they are omitted for clarity. Hydrogen atoms are also omitted to avoid complexity

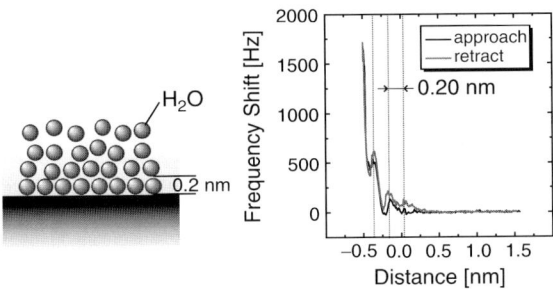

Fig. 13.3. Frequency shift-distance curves measured on the *bc* plane of the *poly*-PTS single crystal in water. Two curves are measured in the approaching and retracting processes, respectively. Left figure shows a schematic model of a hydration structure on a solid surface

13.3 Technical Issues

Achievements of true atomic resolution imaging in liquid by dynamic mode AFM open up a wide variety of application fields. In particular, it is remarkably a powerful tool for the study of molecular scale biology. Furthermore, the applications of various analytical methods based on FM-AFM such as KFM to biology is also greatly expected. On the other hand, there are still several problems to be solved for practical imaging applications.

13.3.1 Force Sensitivity Improvement

Minimum detectable force in FM-AFM is theoretically determined by the Q-factor. An increase of the Q-factor is essentially important for the improvement of the force sensitivity. A major cause for a low Q-factor is the resistance of liquid. The development of novel cantilevers and/or force-sensing devices for liquid imaging, e.g., cantilevers shaped for small liquid resistance or force-sensing quartz resonators having a larger mass than an effective mass of liquid, is required.

13.3.2 Spurious Peaks in Oscillation Spectrum

When the cantilever in liquid is mechanically vibrated, the whole liquid container is also vibrated through liquid. This causes unwanted, various vibrations of the mechanical parts in the proximity of the container, which heavily distorts the cantilever vibration [13]. As a result the oscillation spectrum of the cantilever contains many spurious peaks, which makes not only the identification of resonance of the cantilever more difficult but also the force sensitivity drastically decreased. In this case dynamic mode AFM does not work properly. Thus in order to avoid such problems, "development of a liquid container without generating spurious peaks" and "development of direct excitation method for cantilever vibration" are required. The vibration excitation methods using external magnetic force, photothermal effect or a piezoelectric cantilever are appropriate as excitation techniques producing relatively less spurious peaks.

13.3.3 High-Resolution Imaging of a Sample Having Large Height Variations

In FM-AFM imaging of a sample having high height difference in topography, the cantilever oscillation tends to be unstable or stops in the worst case. When the oscillation stops, the AFM imaging signal is inevitably lost; besides the tip can crash anytime into the sample. By setting the frequency shift at a small value (the tip is positioned far from the sample surface), the risk of the oscillation stop can be eliminated at the severe sacrifice of the imaging resolution. A new intelligent control method of the frequency shift, flexibly changing according to sample topography, is required.

13.4 Roadmap

AFM imaging in liquid has been applied to the studies for biological samples for a long time since the early stage of its development. There has been a tremendous rapid progress in AFM imaging techniques in liquid such as Q-control method and high-resolution imaging using FM–AFM. Figure 13.4 shows a technology roadmap for AFM measurement in liquid, which is based on possible technological developments as well as users' demand. Each subject in the roadmap other than "Developments of cantilevers with a high-Q-factor" and "High-resolution imaging of samples with large height variations", which were explained in Sect. 13.3, is described later.

13.4.1 High-Speed FM-AFM Imaging

There has been a remarkable technological development in high-speed dynamic mode AFM imaging in liquid. High-speed AM-AFM imaging with a frame rate of more than 10 frames per second was already achieved in liquid environments (see Chap. 14) [14]. Imaging speed of FM-AFM is also expected improved. Since it requires a development of an FM detector with a large bandwidth, there are some technological issues to overcome in contrast to AM-AFM. However, a frame rate on the order of $1\,\mathrm{frame\,s^{-1}}$ in FM-AFM is expected achievable in the near future.

13.4.2 Charge Density Mapping in Liquid

Interactions between biological molecules are often based on electrostatic forces. Local charge density mapping of biological molecules in liquid can

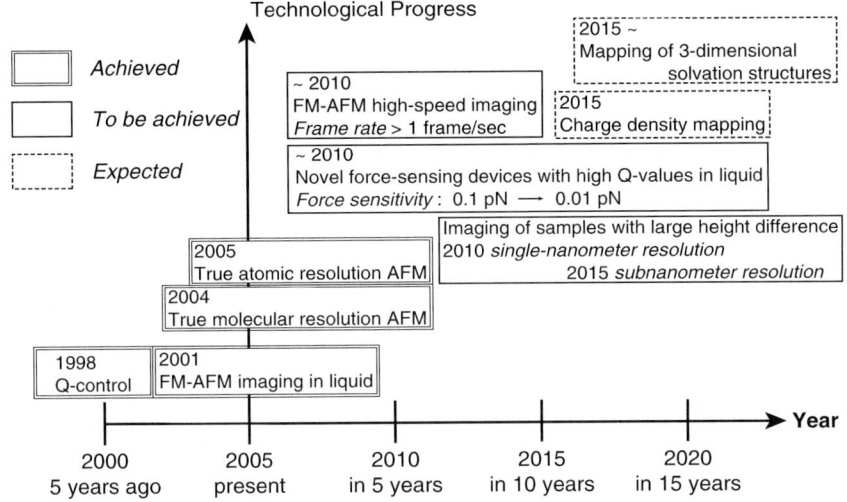

Fig. 13.4. Technology roadmap for AFM measurement in liquid

be a powerful tool for the study of biological functions. While charge density mapping with nanometer resolution is performed in UHV by EFM/KFM, the measurement cannot be simply conducted in liquid because of the existence of charged ions, which can be varied depending on the electrostatic potentials of an AFM tip and a sample. The electrical conductive part of the tip must be shielded except for the tip apex. In addition, the understanding of the potential mapping data obtained requires a close examination of electrolyte double layer effects. However, earlier-mentioned technical problems can be solved by the application of various MEMS techniques as well as advance in liquid AFM technology.

13.4.3 Mapping of Three-Dimensional Solvation Structure

Water plays an essential role in the interactions between biological molecules as well as the structural stability of protein molecules. The hydration structure formed in the proximity of a biological molecule directly affects the intermolecular and/or intramolecular interactions. These solvation/hydration structures have been studied using surface force apparatus and neutron diffraction analysis. Since the solvation shell structures of octamethylcyclotetrasiloxane (OMCTS) were successfully studied recently by AFM force curve measurements, the studies on hydration structures have been made by AFM. A possible, remarkable advantage in AFM study is that it allows us to investigate local solvation structures at a certain position of the molecule. Because the solvation/hydration structure measurement strongly depends on the shape and the chemical property of the AFM tip, the preparation of tips suitable for the solvation/hydration study is essential. Further developments in the measurement resolution and sensitivity of the liquid AFM will probably expand the present AFM study of solvation structures at a single position into a novel mapping method of three-dimensional solvation structures.

References

1. P.K. Hansma et al., Appl. Phys. Lett. **64**, 1738 (1994)
2. R. García, Á. San Paulo, Phys. Rev. B **60**, 4961 (1999)
3. Y. Martin, C.C. Williams, H.K. Wickramasinghe, J. Appl. Phys. **61**, 4723 (1987)
4. R.D. Jäggi et al., Appl. Phys. Lett. **79**, 135 (2001)
5. B. Anczykowski et al., Appl. Phys. A **66**, S885 (1998)
6. Á. San Paulo, R. García, Phys. Rev. B **64**, 193411 (2001)
7. T.R. Rodriguez, R. García, Appl. Phys. Lett. **82**, 4821 (2003)
8. A.D.L. Humphris, J. Tamayo, M.J. Miles, Langmuir **16**, 7891 (2000)
9. J. Tamayo, M.J. Miles, Langmuir **16**, 7891 (2000)
10. K. Kobayashi, H. Yamada, K. Matsushige, Appl. Surf. Sci. **188**, 430 (2002)
11. T. Fukuma et al., Appl. Phys. Lett. **86**, 034103-1 (2005)
12. T. Fukuma et al., Appl. Phys. Lett. **86**, 193108-1 (2005)
13. M. Lantz et al., Surf. Interface Anal. **27**, 354 (1999)
14. T. Ando et al., Proc. Natl. Acad. Sci. **98**, 12468 (2001)

14

High-Speed SPM

Toshio Ando

With a commercially available SPM, it takes a long time, usually a few minutes, to get an image. The limited imaging rate not only imposes on SPM users' patience but also restricts the examinable area size and the scope of examinable dynamics of the sample. Therefore, a high-speed imaging rate has been the capacity of SPM that has been longed for but difficult to realize. Its realization will make it possible to image quickly a large area inaccessible with one shot by scanning its divided areas in rapid succession. This is very effective for evaluating large objects such as semiconductor wafers and liquid crystal panels. In addition, it will become possible to capture the dynamic behavior of the sample in real time. In particular, a dream in life science, i.e., capturing biological macromolecules at work on video, will come true. A high-speed scanning is also an important issue in the application of SPM to nanolithography and atomic-level surface processing. In the practical use of such processed surfaces, the surface area to be processed must be wide. It is impossible to process a large surface area without the SPM's capacity of high-speed scan.

14.1 Optimization of AFM Devices for High-Speed Scanning

For any type of SPM, control of the probe–sample distance is required during scanning. Therefore, in addition to the scanner that does not vibrate even when scanned very fast, high-speed feedback control of the distance is necessary for the realization of high-speed SPM. In the AFM feedback loop, various devices, such as the cantilever, the scanner, the cantilever's deflection detector, the feedback circuit, and the piezo-drive amplifier are involved. All these devices have to be optimized for high-speed scanning. Dividing these devices into three groups, their present capacities and potential for the future developments are summarized below.

14.1.1 Scanner and Related Devices

The sample stage scanner is a mechanical device with a macroscopic size. Therefore, it has a low resonant frequency and thus, is most difficult to optimize for high-speed scanning. To avoid generation of unwanted vibrations its rigidity has to be increased to enhance the resonant frequency. However, in principle it is impossible. To solve this problem a counterbalancing method has been developed [1]. In addition to a piezo actuator that is used for moving the sample stage in the z-direction, another piezo actuator is placed in the opposite direction. These two actuators are displaced simultaneously in the counter direction so as to counteract the impulsive forces caused by their quick displacement. This method can also be used for the x-scanner, and is very effective for reducing vibrations originating in the scanner structure. However, it does not work for reducing the resonant vibrations of the piezo actuators themselves. Active damping of resonant systems has been used widely. When this is applied to the piezo actuators, their displacements should be monitored. But it is difficult in practice. Thus, a new active damping method has been devised; mock piezo actuators are constructed by LCR circuits with the same resonant properties as those of the corresponding piezo actuators, and their output signals are monitored [2]. This method can eliminate the resonant vibrations of the piezo actuators. Because their quality factors are reduced by the damping, their response speed is greatly increased. It is also important to use piezo actuators with a high resonant frequency. However, available piezo materials are limited, and therefore the resonant frequency is inevitably determined almost by the required maximum displacement. Yet, recent development of a technique to stack thin piezo-ceramics films ($\sim 20\,\mu$m) is changing this situation.

Instead of using the sample stage scanner, a self-actuating cantilever coated with ZnO film can be used for z-scan [3]. Cantilevers can be fabricated to small dimensions, and therefore in principle it is easier to enhance their resonant frequency than that of the sample stage scanner. However, because complex fabrication procedures are required to produce self-actuating cantilevers, their resonant frequency is not high enough at present. In addition, their spring constant is large. When a self-actuating cantilever is used in liquid, electrical insulation have to be provided with the cantilever, which requires more complex fabrication processes and hence makes it less possible to afford a high resonant frequency and a small spring constant. These situations will be improved by the progress of MEMS technology in the near future.

14.1.2 Cantilevers and Related Devices

Small cantilevers with a high resonant frequency and a small spring constant have been developed. For instance, Olympus has developed cantilevers with dimensions of 7–9 μm long, 2 μm wide, and about 90 nm thick [4]. Their resonant frequency reaches $>2\,\mathrm{MHz}$ in air and $>1\,\mathrm{MHz}$ in water, and the spring constant is about $0.2\,\mathrm{N\,m^{-1}}$. Although smaller cantilevers have been developed, the method to measure their deflection has become an issue. In addition to the optical lever method, the laser Doppler interferometry has been used for small cantilevers [5]. Although miniaturization of self-sensing cantilevers (using the stress-dependence of the piezo resistor [6]) has not progressed, it will be made possible along with the development of MEMS technology. The electrical circuit for the cantilever's deflection detection sensor with a bandwidth of 20–50 MHz can easily be made. A high-speed amplitude-detection system, required for the AC mode AFM with small cantilevers, has been developed [1]; the peak and bottom-voltage signals of the sinusoidal wave signals are held with S/H circuits, and their difference is output as the amplitude signal at every half cycle of the sinusoidal signals.

14.1.3 Feedback Control and Related Techniques

Modern and advanced control theories have not practically been applied to the high-speed feedback control in AFM. A theory for compensating the phase delays in the feedback loop has been used to some extent. At present the analog feedback control is faster than the digital feedback control, and hence is currently used in the high-speed AFM. However, this situation will change along with the development of the computer and the peripheral I/O boards. The parameters of the conventional PID feedback circuit cannot automatically be altered during scanning, depending on the sample topography. Therefore, enough feedback gain cannot be applied at the steep hills (especially at the down hill) of the sample. To overcome this problem, "dynamic PID control" has been devised [7]; the feedback gain is automatically altered depending on the magnitude of the error signal. In any feedback control the error signal cannot become completely zero, because of its "running after" feature. Thus, the line-by-line or frame-by-frame "feedforward" control has recently been used [8]. When the sample is not moving with time, this control is very effective. A load of task on the feedback control is lightened by the feedforward control, and therefore, high-speed scanning becomes possible. For the sample moving with time, the feedforward control has a possibility to operate in a wrong direction. However, it becomes more effective with the higher scan rate.

The amplitude modulation (AM) and the frequency modulation (FM) of the cantilever oscillation have been used to get information other than the sample topography. However, in this technique the quality factor (Q) of the cantilever oscillation is increased to enhance its sensitivity to the tip–sample interaction, and hence it cannot be compatible with the high-speed imaging. But, this situation is changing. Because of the high resonant frequency and of the small spring constant of the small cantilevers, their oscillation properties are sensitive to the tip–sample interaction even with a small Q. Therefore, a large frequency shift and a large phase shift are expected to occur with the

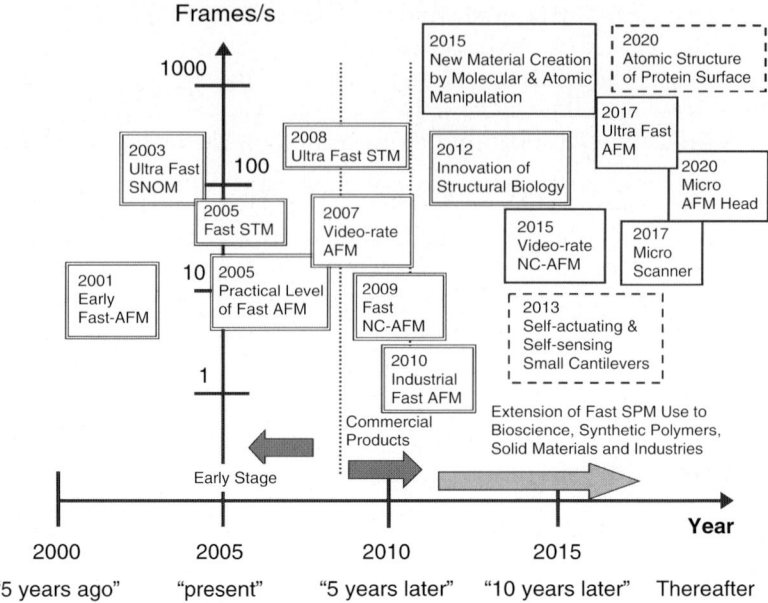

Fig. 14.1. Roadmap of high-speed SPM

small cantilever with a small Q. Accordingly, slow measurement-instruments such as the lock-in amplifier and the PLL circuit are not necessary for the shift measurements, and therefore, there is a possibility that high-speed NC-AFM is realized in the near future (see Fig. 14.1).

14.2 World Trends in the High-Speed SPM

The enhancement of the imaging rate by miniaturizing cantilevers was challenged by the UC Santa Barbara group led by Paul Hansma. The imaging rate was a few seconds/frame in the tapping mode of operation [9]. Recently, they developed a high-speed scanner. Although their latest AFM has a capacity to image a hard sample in air at $\sim 0.1\,\text{s}\,\text{frame}^{-1}$, they have not yet succeeded in the fast imaging of protein molecules in solution. The Kanazawa University group led by Toshio Ando has challenged to optimize all the AFM devices for high-speed scanning, and reached an imaging rate of $60\,\text{ms}\,\text{frame}^{-1}$, which made it possible to capture the dynamic behavior of protein molecules in solution [1, 7] (Fig. 14.2). The group of the University of Bristol led by Mervyn Miles has tried to increase the imaging rate by giving up the feedback operation [10]. They attached a sample stage to a tuning fork that was vibrating at its resonant frequency in the x-direction. Therefore, the fast x-scan is not made with triangular waves but with sinusoidal waves. They have achieved an imaging rate of $30\,\text{ms}\,\text{frame}^{-1}$. Because of no feedback operation, the applicable

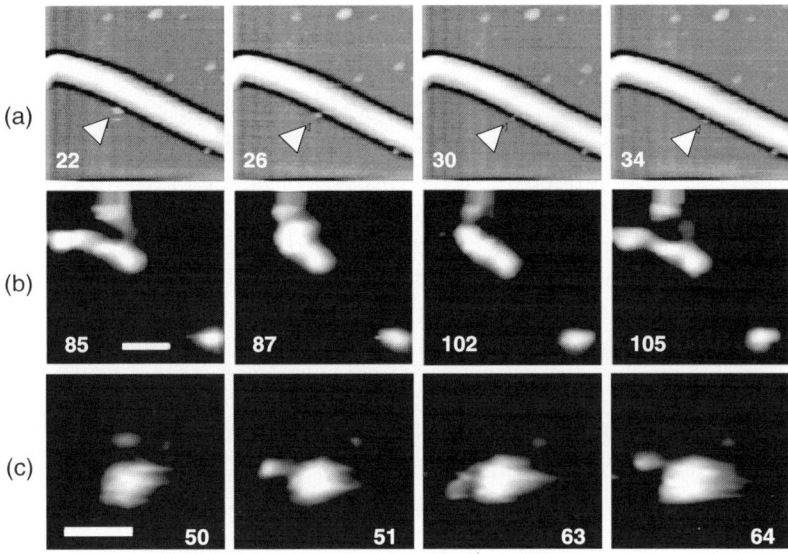

Fig. 14.2. Dynamic behaviors of motor proteins captured by the high-speed AFM. The movies are placed at `www.s.kanazawa-u.ac.jp/phys/biophys/roadmap.htm`. **(a)** Kinesin-gelsolin chimera moving along a microtubule (collaboration with Toyoshima Lab at the University of Tokyo). The chimera protein is moving toward the lower right (*marked with arrows*). The thick rod is a microtubule. The number on each image is the frame number. The frame rate is 640 ms frame^{-1}. Scan size, 400 nm. **(b)** Myosin V before and after releasing ATP from caged-ATP by the irradiation of UV light pulses. Frame-85 is the image just before releasing ATP, and Frame-87 is the image just after releasing ATP. One of the two heads of myosin V bent immediately after the release of ATP. Its bent conformation continued for a while, and then returned to the original straight form. The frame rate is 80 ms frame^{-1}. Scale bar, 30 nm. **(c)** Conformational change in dynein C caused by the ATPase reaction (collaboration with Ooiwa Lab of the Kansai Advanced Research Center). The left-hand protrusion from the ring structure is "stem," and the upper right protrusion from the ring is "stalk." The stem often stayed at the position indicated with Frame-51 and Frame-64, occasionally (once per a few frames) moved to the position indicated with Frame 63 and returned to the previous position. The frame rate is 160 ms frame^{-1}. Scale bar, 30 nm

samples are limited. However, for rigid samples, a wide area of the sample has been imaged quickly by scanning the divided areas in rapid succession. The Stanford University group led by Calvin Quate developed cantilever arrays with the self-actuation capability [11], and succeeded in imaging a wide area of the sample at once. An MIT group developed an AFM with a time resolution of μs [12]. This is not a real-time imaging. An external stimulus (such as a light pulse) is applied to the sample, and the cantilever tip synchronously detects the sample response at one spot and then is moved to a neighboring spot. This procedure is repeated until all the area to be imaged is covered. Since there would not exist microscopic samples that always respond to stimuli in the same timing, application of this AFM seems very limited.

The scanner, the feedback control and the data processing are basically common to any SPM, and their optimization techniques mentioned above for high-speed AFM are applicable to the other types of SPM. Here, let us survey briefly high-speed SPMs (other than AFM) that have been developed so far. A group of the Leiden University has developed a high-speed STM [13]. They have made efforts to enhance the bandwidth of all the electronics involved. In addition, they produced a high-speed scanner with a simple structure of small dimensions; two shear-mode piezo actuators with the maximum displacement of 300 nm are stacked to produce a XY-scanner on the top of which a z-piezo is attached. They succeeded in obtaining atomic-resolution images (the scan range of $\sim$50 nm) at the maximum speed of $\sim$10 ms frame^{-1}. However, the resonant frequency of the small scanner is not high enough, and therefore the x-scan is made with sinusoidal waves. An ultra high-speed NSOM has been developed by the group of the University of Bristol [14]. In the same way as the high-speed AFM they developed, the feedback operation is given up. A vertically aligned optical fiber probe is attached to a tuning fork that is oscillating at the resonant frequency in the x-direction. Using a piezo actuator, the base of the tuning fork is scanned in the y-direction (the slow scan direction). This simple design allows imaging at 8.3 ms frame^{-1}. This NSOM has been used to capture crystallization processes of some materials.

14.3 Roadmap

14.3.1 Until 2010

The basic architecture of SPM will not change for a while. However, the capacity of each device of SPM will be enhanced. It will lead to a video-rate SPM that can control the tip–sample interaction force at a small level, and can capture fragile samples on video without giving damage to the sample. With STM that images atomically flat samples, a frame rate of a few ms frame^{-1} will be achieved. With AFM that observes large samples such as cells, $\sim$1 s frame^{-1} will be achieved. High-speed AFM will certainly come onto the market. However, the main products will not be the one that aims at observing dynamic processes of the sample, but will be intended to enhance the observation efficiency with an imaging rate of $\sim$1 s frame^{-1}. The mass-production techniques for small cantilevers will have been established.

14.3.2 Until 2015

Owing to the progress of MEMS technology, self-actuating/self-sensing small cantilevers will be developed. They have a high resonant frequency and a small spring constant, even when electrically insulated for the use in liquid. The AFM topography-imaging rate will go beyond video-rate, and reach $\sim 10\,\mathrm{ms\,frame^{-1}}$. The rate of mapping the surface physicochemical properties with the AM- and FM-AFM will become closer to video-rate. While the use of high-speed AFM for industries will be expanded, the video-rate AFM will commercially be available and will begin to be used widely in life science.

14.3.3 After 2015

Not only cantilevers but also tiny sample-stage scanners will be fabricated by MEMS techniques. The dimensions of the AFM head will become $\sim 10\,\mathrm{mm^3}$. The scan speed will be dramatically enhanced, and thereby an ultra high-speed AFM with the topography-imaging rate of a few $\mathrm{ms\,frame^{-1}}$ will be born. As a result, observations of dynamic processes at the molecular and atomic levels will be performed in various fields. In the field of biology, the elucidation of protein function will be progressed due to the knowledge of the dynamic behavior. It will be possible to construct and analyze dynamic atomic models by introducing the observed dynamic behaviors to the static atomic models. Thus, structural biology will be innovated. In addition, along with the enhancement of the spatial resolution the protein's secondary structure and atomic arrangements on the surface will be resolved. Owing to a dramatic enhancement of the scan speed, the surface processing at the atomic level which can now be performed only on a small surface area will become possible also on a macroscopic area. Thus, surfaces that are atomically processed will be provided for their practical use. In addition, material surfaces newly created by atomic processing will become the new subject of material researches.

References

1. T. Ando, N. Kodera, E. Takai, D. Maruyama, K. Saito, A. Toda, Proc. Natl. Acad. Sci. USA **98**, 12468 (2001)
2. N. Kodera, H. Yamashita, T. Ando, Rev. Sci. Instrum. **76**, 053708 (2005)
3. T. Itoh, C. Lee, T. Suga, Appl. Phys. Lett. **69**, 2036 (1996)
4. M. Kitazawa, K. Shiotani, A. Toda, Jpn. J. Appl. Phys. **42**, 4844 (2003)
5. H. Kawakatsu, S. Kawai, D. Saya, M. Nagashio, D. Kobayashi, H. Toshiyoshi, H. Fujita, Rev. Sci. Instrum. **73**, 2317 (2002)
6. S.C. Minne, S.R. Manalis, C.F. Quate, Appl. Phys. Lett. **67**, 3918 (1995)
7. T. Ando, N. Kodera, Y. Naito, T. Kinoshita, K. Furuta, Y.Y. Toyoshima, Chem. Phys. Chem. **4**, 1196 (2003)
8. G. Schitter, R.W. Stark, A. Stemmer, Ultramicroscopy **100**, 253 (2004)
9. M.B. Viani, T.E. Schäffer, G.T. Paloczi, L.I. Pietrasanta, B.L. Smith, J.B. Thompson, M. Richter, M. Rief, H.E. Gaub, K.W. Plaxco, A.N. Cleland, H.G. Hansma, P.K. Hansma, Rev. Sci. Instrum. **70**, 4300 (1999)
10. A.D.L. Humphris, M.J. Miles, J.K. Hobbs, Appl. Phys. Lett. **86**, 034106 (2005)

11. S.C. Minne, G. Yaralioglu, S.R. Manalis, J.D. Adams, J. Zesch, A. Atalar, C.F. Quate, Appl. Phys. Lett. **72**, 2340 (1998)
12. M. Anwar, I. Rousso, Appl. Phys. Lett. **86**, 014101 (2005)
13. M.J. Rost, L. Crama, P. Schakel, E. van Tol, G.B.E.M. van Velzen-Williams, C.F. Overgauw, H. ter Horst, H. Dekker, B. Okhuijsen, M. Seynen, A. Vijftigschild, P. Han, A.J. Katan, K. Schoots, R. Schumm, W. van Loo, T.H. Oosterkamp, J.W.M. Frenken, Rev. Sci. Instrum. **76**, 053710 (2005)
14. A.D.L. Humphris, J.K. Hobbs, M.J. Miles, Appl. Phys. Lett. **83**, 6 (2003)

Scanning Nonlinear Dielectric Microscope

Yasuo Cho

15.1 Principle and Theory for SNDM

Recently, ferroelectric materials have attracted the attention of many researchers. Their large dielectric constants make them suitable as dielectric
layers of microcapacitors in microelectronics. They are also investigated for
application in nonvolatile memory using the switchable dielectric polarization of ferroelectric material. To characterize such ferroelectric materials, a
high-resolution tool is required for observing the microscopic distribution of
remanent (or spontaneous) polarization of ferroelectric materials.

With this background, we have proposed and developed a new purely electrical method for imaging the state of the polarizations in ferroelectric and
piezoelectric material and their crystal anisotropy. It involves the measurement of point-to-point variations of the nonlinear dielectric constant of a specimen and is termed "scanning nonlinear dielectric microscopy (SNDM)" [1–5].
This is the first successful purely electrical method for observing the ferroelectric polarization distribution without the influence of the screening effect from
free charges. To date, the resolution of this microscope has been improved
down to the subnanometer order.

Figure 15.1 shows the system setup of the SNDM using the LC lumped
constant resonator probe. In the figure, C_s denotes the capacitance of the
specimen under the center conductor (the tip) of the probe. C_s is a function of
time because of the nonlinear dielectric response under an applied alternating
electric field $E_{p3}(= E_p \cos \omega_p t, \ f_p = 5 - 200\,\text{kHz})$.

This LC resonator is connected to the oscillator tuned to the resonance
frequency of the resonator. The earlier mentioned electrical parts (i.e., tip
(needle or cantilever), ring, inductance and oscillator) are assembled into a
small probe for the SNDM. The oscillating frequency of the probe (or oscillator) (around 1.3–2 GHz) is modulated by the change of capacitance $\Delta C_s(t)$
due to the nonlinear dielectric response under the applied electric field. As
a result, the probe (oscillator) produces a frequency modulated (FM) signal.
By detecting this FM signal using the FM demodulator and lock-in amplifier,

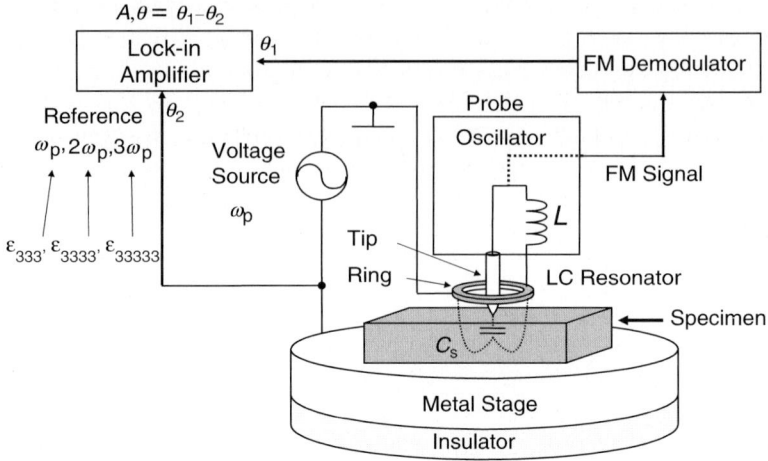

Fig. 15.1. Schematic diagram of SNDM

we obtain a voltage signal proportional to the capacitance variation. Thus we can detect the nonlinear dielectric constant just under the needle and can obtain the fine resolution determined by the diameter of the pointed end of the tip and the linear dielectric constant of specimens. The capacitance variation caused by the nonlinear dielectric response is quite small ($\Delta C_{\rm s}(t)/C_{\rm s0}$ is in the range from 10^{-3} to 10^{-8}). Therefore the sensitivity of SNDM probe must be very high. The measured value of the sensitivity of an earlier mentioned lumped constant probe is 10^{-22} F which is much higher than that of scanning capacitance microscope (SCM) whose typical sensitivity is 10^{-18} F.

15.2 Microscopic Observation of Area Distribution of Ferroelectric Domain Using SNDM

Originally, this microscope has been developed for the purpose of measuring the ferroelectric polarization and the local anisotropy of dielectrics through the detection of third-rank nonlinear dielectric constant. In Fig. 15.1, the capacitance immediately beneath the tip $C_{\rm s}$ is modulated alternately due to the nonlinear dielectric properties of the specimen cause by the application of alternating electric field between the tip and the metal stage. The ratio of this derivative capacitance variation $\Delta C_{\rm s}$ to the static value to the capacitance $C_{\rm s0}$ is given by (15.1).

$$\frac{\Delta C_{\rm s}}{C_{\rm s0}} = \frac{\varepsilon(3)}{\varepsilon(2)} E_{\rm p}\cos(\omega_{\rm p}t) + \frac{1}{4}\frac{\varepsilon(4)}{\varepsilon(2)} E_{\rm p}^2\cos(2\omega_{\rm p}t) + \frac{1}{24}\frac{\varepsilon(5)}{\varepsilon(2)} E_{\rm p}^3\cos(3\omega_{\rm p}t) + \cdots,$$

$$(15.1)$$

where $\varepsilon(2)$, $\varepsilon(3)$, $\varepsilon(4)$, and $\varepsilon(5)$ are a linear, lowest order nonlinear, one order-higher, and two-order higher nonlinear dielectric constant, respectively. (The number in the bracket denotes the rank of tensor.)

This equation shows that the alternating capacitance of different frequencies corresponds to each order of the nonlinear dielectric constant. Signals corresponding to $\varepsilon(3)$, $\varepsilon(4)$, and $\varepsilon(5)$ were obtained by setting the reference signal of the lock-in amplifier in Fig. 15.1 to frequency ω_p, $2\omega_p$, and $3\omega_p$ of the applied electric field, respectively.

The even rank tensors, including the linear dielectric constant $\varepsilon(2)$, are insensitive to the states of the spontaneous polarization. On the other hand, the lowest-order nonlinear dielectric constant $\varepsilon(3)$ and other higher order odd rank tensor nonlinear dielectric constant are very sensitive to spontaneous polarization. For example, there is no $\varepsilon(3)$ in a material with a center of symmetry, the sign of $\varepsilon(3)$ changes in accordance with the inversion of the spontaneous polarization. Therefore, by detecting this $\varepsilon(3)$ microscopically, we can measure the area distribution of ferroelectric domain.

As one example of domain measurement, here, we show the $\varepsilon(3)$ images in Fig. 15.2 which demonstrates the resolution of SNDM is really subnanometer order. The tip used in this measurement was a metal coated conductive cantilever with tip radius of 25 nm.

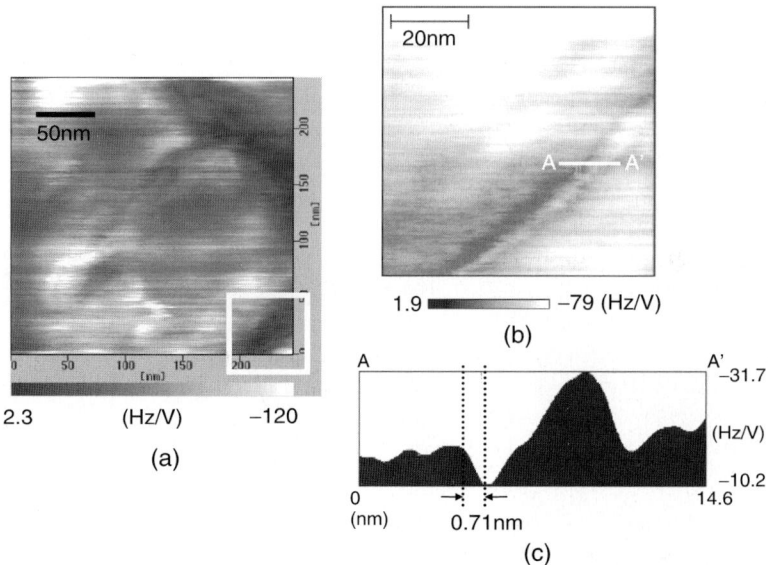

Fig. 15.2. Domain pattern taken from an epitaxial PZT thin film on La–Sr–Co–O/SrTiO$_3$ (nonliear dielectric $\varepsilon(3)$ image) (**a**) Image of 90° a–c domain, (**b**) Magnified image of the area surrounded by the white square, (**c**) Cross-sectional image taken along A–A′

These images were taken from a epitaxial lead zirconate titanate (PZT) thin (4,000 Å)/La–Sr–Co–O/SrTiO₃. The stripe shape 90° a–c domain pattern is seen in Fig. 15.2a. Figure 15.2b is the magnified image of the area surrounded by the white square in Fig. 15.2a. Figure 15.2c is a cross-sectional image taken along line A–A′ in Fig. 15.2b. From the distance between the clearly distinguishable structures in the image, it is apparent that SNDM has subnanometer resolution.

15.3 Visualization of Stored Charge in Semiconductors Using SNDM

Because the SNDM is capable of detecting an ac capacitance change of 10^{-22} F, which is extraordinarily small, the methods described earlier can be used not only to detect ferroelectric polarization, but also in any applications that involve samples containing dipoles. As an illustrative example of investigating systems other than ferroelectrics, in Fig. 15.3 we show the results of our experimental visualizations of electrons and holes stored in metal oxide nitride oxide semiconductor (MONOS) type flash memories [6]. We were able to visualize the dipoles created when an electron (hole) stored in the oxide nitride oxide (ONO) layer induces a hole (electron) at the silicon surface, which amounts to detecting the negatively charged electrons and the positively charged holes.

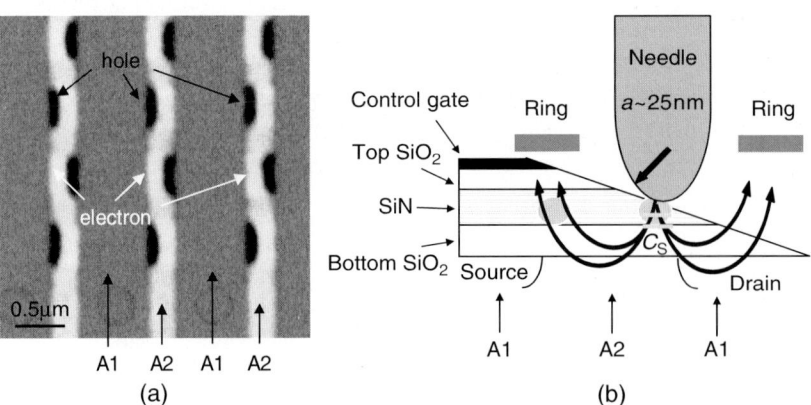

Fig. 15.3. Visualization of electrons and holes localized in the thin gate film of MONOS type semiconductor flash memory. (**a**) SNDM image of sample into which both electron and hole injected, creating a checked pattern. (**b**) Schematic cross-section of the sample

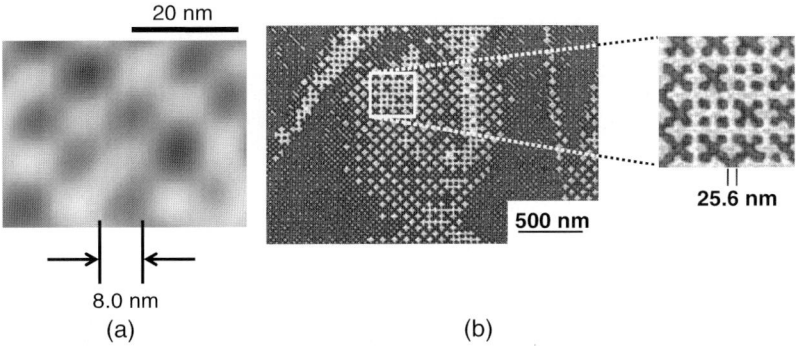

Fig. 15.4. SNDM ferroelectric probe memory: (**a**) areal memory density of $10\,\text{Tbit\,in.}^{-2}$ (close packed nanodomain dots); (**b**) areal memory density of $1\,\text{Tbit\,in.}^{-2}$ (actual information storage)

15.4 SNDM Ferroelectric Probe Memory

With the advance of information processing technology, the importance of high-density data storage is increasing. Studies on thermal fluctuation predict that magnetic storage, which plays a major role in this field, will reach a theoretical limit in the near future, and thus a novel high-density storage method is required.

Ferroelectrics can hold bit information in the form of the polarization direction of individual domains. Moreover, the domain wall of typical ferroelectric materials is as thin as the order of a few lattices, which is favorable for high-density data storage. Therefore, we have been studying ferroelectric high-density data storage based on scanning nonlinear dielectric microscopy (SNDM) [7]. Thus, as shown in Fig. 15.4, nanosized inverted domain dots have been successfully formed at a data density of $10\,\text{Tbit\,in.}^{-2}$ and actual information storage is demonstrated at a density of $1\,\text{Tbit\,in.}^{-2}$

15.5 Roadmap

Figure 15.5 shows the future prediction roadmap for SNDM. As shown in the figure, this technique has started from the detection of simple perpendicular component of ferroelectric polarization to the surface and now its coverage area is being rapidly expanded.

For example Ca higher order nonlinear dielectric microscopy (HO-SNDM) technique [8] with higher lateral and depth resolution than conventional nonlinear dielectric imaging and a noncontact scanning nonlinear dielectric microscopy (NC-SNDM) which focuses on the nonlinear dielectric signal for controlling the noncontact condition have been developed [9]. With NC-SNDM technique, topographic and SNDM images were obtained simultane-

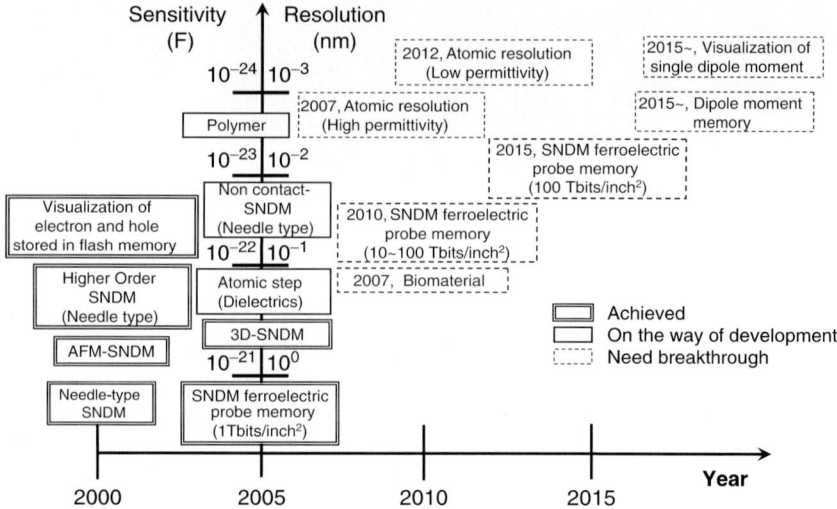

Fig. 15.5. The roadmap of scanning nonlinear dielectric microscopy

ously. Moreover, from the calculation results, NC-SNDM has been expected to have an atomic resolution.

Moreover, 3D-SNDM technique has also been developed for both vertical and lateral polarization measurement [10].

Next challenging themes to be solved up to 2015 are achievement of atomic scale observation of the insulator surface and the local polarizability measurement of polar material with atomic scale resolution. Moreover, SNDM ferroelectric probe memory with the areal memory density above 100 Tbit in.$^{-2}$ can be fully expected after 2015.

References

1. Y. Cho, A. Kirihara, T. Saeki, Denshi Joho Tsushin Gakkai Ronbunshi (in Japanese) **J78-C-1**, 593 (1995)
2. Y. Cho, A. Kirihara, T. Saeki, Rev. Sci. Instrum. **67**, 2297 (1996)
3. Y.Cho, S. Atsumi, K. Nakamura, Jpn. J. Appl. Phys. **36**, 3152 (1997)
4. Y. Cho, S. Kazuta, K. Matsuura, Appl. Phys. Lett. **75**, 2833 (1999)
5. H. Odagawa, Y. Cho, Surf. Sci. **463**, L621 (2000)
6. K. Honda, S. Hashimoto, Y. Cho, Nanotechnology **16**, S90 (2005)
7. Y. Cho, S. Hashimoto, N. Odagawa, K. Tanaka, Y. Hiranaga, Appl. Phys. Lett. **87**, 232906 (2005)
8. Y. Cho, K. Ohara, Appl. Phys. Lett. **79**, 3842 (2001)
9. K. Ohara, Y. Cho, Nanotechnology **16**, S54 (2005)
10. T. Sugihara, H. Odagawa, Y. Cho, Jpn. J. Appl. Phys. **44**, 4325 (2005)

SPM Coupled with External Fields

Ken Nakajima and Tadahiro Komeda

In this section, the novel techniques coupling scanning probe microscope (SPM) with external field excitations are described. The emphasis will be placed on their present states of arts together with their future prospects.

16.1 Light-Illumination STM

The most representative example of the combined technology between SPM and an external field is light-illumination scanning tunneling microscope (STM), where the tunneling current modulated by light-illumination becomes the measure. It is expected that the method will be applicable to widely spreading scientific fields since the method provides the simultaneous detection of the optical information on structural or electronic transition by light excitation with the conventional high-resolution STM feature. The road map for light-illumination STM is summarized in Fig. 16.1. Further detailed information is also available in a textbook [1].

The history of this technique is not short. The basic concept had been reported early in the 1990s [2]. The system is commonly composed of STM with a metallic probe and far-field optics where the laser beam is irradiated to the tunneling gap from the side. However, there has been no distinct consensus on its principle; how nanometer-sized tunneling gap beneath the STM probe can be optically excited? Although the near-field excitation accounts for the phenomena as a possible tentative agreement, further studies are indispensable to elucidate the point using electromagnetic field analysis [3]. Another practical and serious problem exists in the thermal expansion of probe or sample itself by light illumination [4]. A sufficiently fast mechanical chopping technique is thus cooperated to avoid the problem. Alternatively, the light intensity must be reduced. These constraints are not preferable when the applicability of this technique becomes a concern. Since the light illumination is made on a larger area comparing with tunneling gap itself, an optical damage sometimes becomes more serious. The question also arises that the observation can be really caused by any local phenomenon. A technique combining STM with near-field

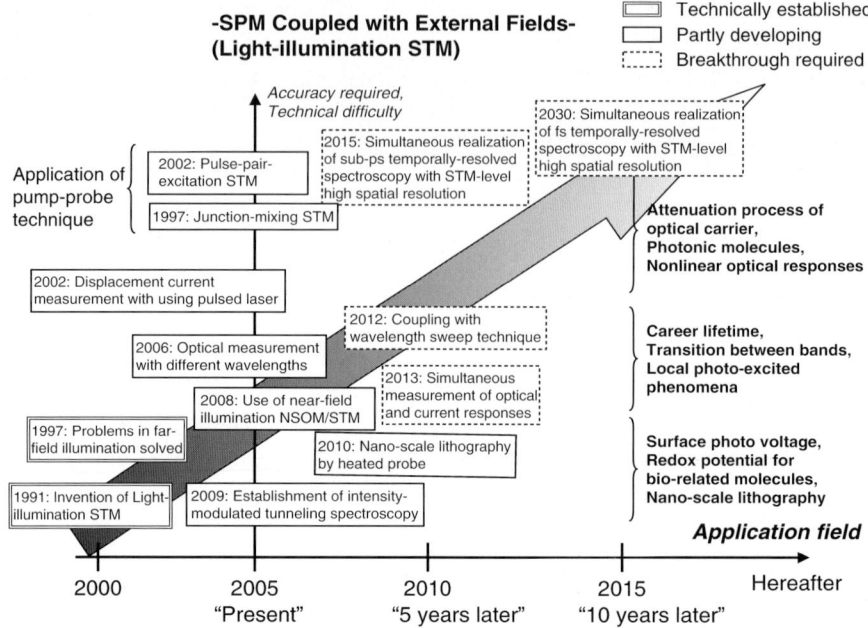

Fig. 16.1. The road map of light-illumination STM

scanning optical microscope (NSOM) is proved to be useful to overcome these problems to some extent [5], while the technique needs more sophistication to be a general-purpose use.

Though there are several problems as mentioned earlier, it is a fact that there have already been opened so many application examples as shown in Fig. 16.1. The followings are the apt illustrations; measurements on surface photo voltage of semiconductor [6], carrier lifetime [7], and the application to nanometer-scale lithography using heated probe [8]. The future prospective researches may include nonlinear optical responses, the combination with wavelength sweep techniques in order to expand an applied field and market. In addition, recent developments on the combination with ultra short pulse laser is in the process of realizing high temporal resolution that conventional STM is difficult to do. The novel techniques such as junction-mixing STM [9] and pulse-pair-excited STM [10] will unveil surface fast dynamics in the near future. The progress within the coming ten years will definitely be a key to realize it.

16.2 Coupling with Outer Field; Electron Spin Resonance Detection using STM

There have been continuous efforts in STM studies for the detection of spins on the surface. The detection methods can be classified into the following three groups. (1) Detection of spin-sensitive density-of-states by STS using spin polarized tip, (2) detection of Zeeman splitting by measuring inelastic channel of the tunneling current which flip the spin direction, (3) detection of Larmor precession in the presence of a magnetic field which appears as a periodic noise in the tunneling current.

With the use of method (3), it might be possible to measure the g value which is currently used for chemical analysis in ESR measurement. Moreover, in addition to the detection of the presence of spins, the determination of spin direction is of great importance for an elemental process of quantum computing which might be measured with this technique.

Thus let us discuss the mechanism of the method (3) (Fig. 16.2a). We assume an isolated spin S and tunneling electrons which are polarized to a certain direction relative direction makes a difference in tunneling barrier which can be described as $J(S \cdot \sigma)$. J represents the Heisenberg exchange energy [11, 12]. Assume we have one-dimensional square lattice potential. The tunneling current is proportional to $\exp(-\sqrt{\phi/\phi_0})$. Here ϕ represents the tunneling barrier height, $\phi_0 = \hbar/8md^2$, where m is the mass of electron and d the gap distance of tunneling barrier.

Since the barrier varies with the isolated spin and the tunneling electrons, we replace $\phi \rightarrow \phi - J(s \cdot \sigma)$. Then the current is expressed as $I = \exp(-\sqrt{(\phi - JS\sigma)/\phi_0})$. We separate I into spin-dependent part ΔI and spin independent part I_0. $\Delta I = \exp(-\sqrt{\phi/\phi_0}) \sqrt{\phi/\phi_0} (JS\sigma)/(2\phi)$ total current is a sum of dc component and sine-wave of spin dependent component. Spin dependent component and the frequency is expressed as $\omega_L = g\mu_B B/\hbar$. The ratio of the two components can be expressed as $\langle \Delta I^2 \rangle^{1/2}/I_0 = 2/\sqrt{N} \sqrt{\phi/\phi_0} JS/(2\phi)$. Here N represents the number of tunneling electrons for the time period of one-cycle of Larmor precession. In case N is infinite the ratio converge into zero. However, the limited number of tunneling electrons this ratio survives. ΔI is detectable with the set up of STM. If we input normal values for the parameters, then we expect that ~1% of the total current contributes to the spin signal. This much of the current is in the detectable range.

This technique has been pioneered by Manassen and coworkers which detected the spin in the Si oxide originated from dangling bond for which the presence of the spin center is widely accepted [13]. But Manassen has demonstrated for the first time that the tunneling current is modulated with Larmor frequency.

More recently Durkan et al. showed that a molecule of α,γ-bisdiphenylene-β-phenylallyl adsorbed on HOPG surface also shows a clear spectrum [14] (Fig. 16.2b).

Expected future development is illustrated in Fig. 16.2c. Though clear evidence of the detection of Larmor frequency component in the tunneling current has been obtained, all the experiments were conducted at room temperature and it is not convincing that a single spin is actually detected. In addition the tip used in these experiments was not spin polarized, and the evaluation

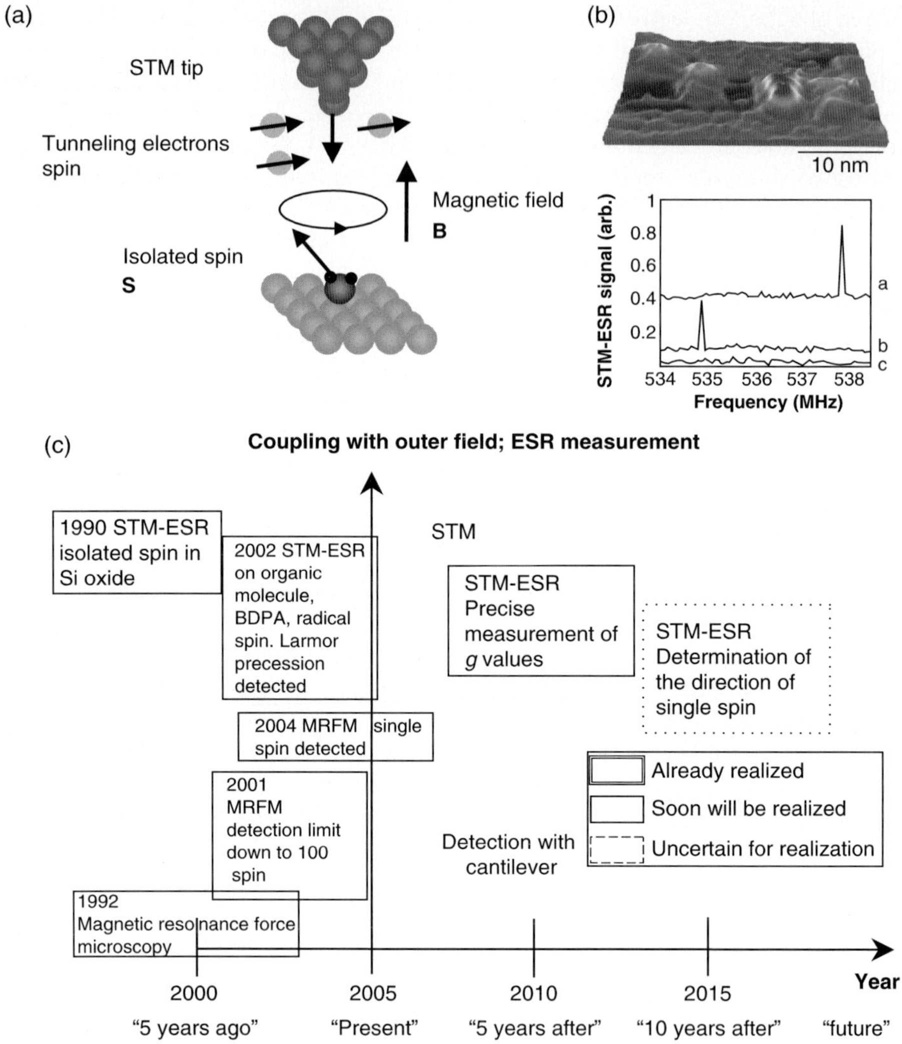

(a)

STM tip

Tunneling electrons
spin

Magnetic field
B

Isolated spin
S

(b)

10 nm

STM-ESR signal (arb.)

534 535 536 537 538
Frequency (MHz)

a
b
c

(c)   **Coupling with outer field; ESR measurement**

STM

1990 STM-ESR
isolated spin in
Si oxide

2002 STM-ESR
on organic
molecule,
BDPA, radical
spin. Larmor
precession
detected

STM-ESR
Precise
measurement of
g values

STM-ESR
Determination of
the direction of
single spin

2004 MRFM single
spin detected

2001
MRFM
detection limit
down to 100
spin

Detection with
cantilever

☐ Already realized

☐ Soon will be realized

⊡ Uncertain for realization

1992
Magnetic resonance force
microscopy

2000   2005   2010   2015   **Year**

"5 years ago"   "Present"   "5 years after"   "10 years after"   "future"

Fig. 16.2. (a) Principle of operation of detecting ESR signal by STM. Relation between the precession of isolated spin S and spin σ of tunneling electron in magnetic field B, **(b)** a molecule of α, γ-bisdiphenylene-β-phenylallyl adsorbed on HOPG surface also shows a clear spectrum [14], **(c)** roadmap of STM–ESR

indicates about 1% of the dc component appear in the AC component. However, the observed AC component is in order of magnitude higher than the theoretical prediction. Apparently there needs much more accumulation of data both in theoretical and experiment. To elucidate much more solid physical principle of this technique experiments in ultra high vacuum condition and low temperature is necessary.

16.3 Roadmap

As described earlier, two major attempts to couple SPM with external field excitations are explained in terms of their present states of arts and future prospects. STM–ESR will become a powerful tool to detect a single spin property together with another important technique, magnetic resonance force microscope (MRFM). These two techniques will be unified in the future.

References

1. R. Wiesendanger, *Scanning Probe Microscopy and Spectroscopy – Methods and Applications*, (Cambridge University Press, Cambridge, 1994)
2. M. Völcker, W. Kreiger, H. Walther, Phys. Rev. Lett. **66**, 1717 (1991)
3. O.J.F. Martin, C. Girald, Appl. Phys. Lett. **70**, 705 (1997)
4. I. Lyubinetsky, Z. Dohnálek, V.A. Ukraintsev, J.T. Yates Jr., J. Appl. Phys. **82**, 4115 (1997)
5. K. Nakajima, B. Lee, S. Takeda, J. Noh, T. Nagamune, M. Hara, Jpn. J. Appl. Phys. **42**, 4861 (2003)
6. S. Grafström, J. Appl. Phys. **91**, 1717 (2002)
7. R.J. Hamers, D.G. Cahill, J. Vac. Sci. Technol. B **9**, 514 (1991)
8. H. Shigekawa, J. Surf. Sci. Soc. Jpn. (in Japanese) **20**(5), 336 (1999)
9. M.R. Freeman, A.Y. Elezzabi, G.M. Steeves, G. Nunes Jr., Surf. Sci. **386**, 290 (1997)
10. O. Takeuchi, R. Morita, M. Yamashita, H. Shigekawa, Jpn. J. Appl. Phys. **41**, 4994 (2002)
11. A.V. Balatsky, Y. Manassen, R. Salem, Phys. Rev. B **66**, 1954161 (2002)
12. A.V. Balatsky, Y. Manassen, R. Salem, Philos. Mag. A **82**, 1291 (2002)
13. Y. Manassen, R.J. Hamers, J.E. Demuth, A.J. Castellano, Phys. Rev. Lett. **62**, 2531 (1989)
14. C. Durkan, A. Ilie, M.S.M. Saifullah, M.E. Welland, Appl. Phys. Lett. **80**, 4244 (2002)

Probe Technology

Masamichi Yoshimura

17.1 Introduction

The shape and mechanical stability of probe tips are crucial in scanning probe microscopy (SPM), since they determine the spatial resolution of SPM. If the size of the tip apex is larger than the object to be observed, the correct image of the object is hard to obtain. If the tip apex has several protrusions, obtained images show so called ghost structure due to the multiple tip effect. You should be careful especially when you count the density of nanoclusters or nanoparticles. Thus, the size of tip apex should be sharp and small enough as compared with the object to be measured.

Now, we consider the whole part of the probe rather than the apex. If the sample has a deep hole like a trench in semiconductor devices, it is not so easy to get a complete figure of the hole. In order to measure correctly the hole, the aspect ratio of the probe should be smaller than that of the hole. Multiprobe SPM is now a hot topic, where two or four probes are used simultaneously to measure electric conductance of nanomaterials such as nanowires. In this case, the geometrical interference between adjacent probes is a big problem to be solved. In these cases, a small probe with high-aspect ratio is highly required.

As scanning tunneling microscopy (STM) probes, metal wires such as tungsten have been generally used. They are electrochemically or mechanically sharpened in many ways. In ultra-high vacuum (UHV), a tungsten tip is widely used and is cleaned and sharpened by heating, sputtering, and faceting before use, which leads to high spatial resolution and mechanical stability. For use in air, Pt-alloy materials (difficult to oxidize) are widely used. In solution, glass or wax is coated over the metal wires to reduce dark current. Thus, the STM probes can be easily fabricated in the laboratory.

For atomic force microscopy (AFM), many kinds of probes are available depending on the type of measurement, and most of them are based on cantilever technology. It is rather difficult to fabricate in the laboratory, but commercial products with variable specification parameters such as resonant oscillation, Q-value are available. It should be kept in mind that the probe apex will wear after use. The test samples for evaluating the tip shape are also available commercially.

Two kinds of probes, scattering type and aperture type, are used in scanning nearfield optical microscopy (SNOM) depending on the type of measurement. In the former, metal, semiconductor, and dielectric materials are commonly used. In the latter, sharpened optical fiber by etching and so on is generally used. They are coated with metal films and the aperture is fabricated by focus ion beam (FIB) or by breaking the apex in some ways.

17.2 Carbon Nanotube Probe

Carbon nanotube (CNT) is one of intriguing materials in nanotechnology. CNT has a one-dimensional tube structure with unique physical characteristics such as small diameter, high-aspect ratio, high stiffness, high conductivity. These properties are best fit to the tip in SPM probe. Several methods to fabricate CNT–SPM probe, namely, to attach CNT on the conventional probe, have been reported. They are classified into dry and wet methods. The former is further divided into mechanical and chemical methods. In the latter, dielectrophoresis technique is widely conducted.

17.2.1 Mechanical Method

The first research to fabricate CNT probe has been conducted under optical microscope [1]. They fixed a CNT to the epoxy-attached probe using a precise mechanical stage. Then, the attachment of CNT onto the probe has been performed using two independent precise stages under SEM observation (Fig. 17.1), where beam-induced deposition of amorphous carbon is used as glue [2]. The length of CNT can be controlled [3], and the mechanically fabricated CNT probes are used for the lithography [4] and magnetic force

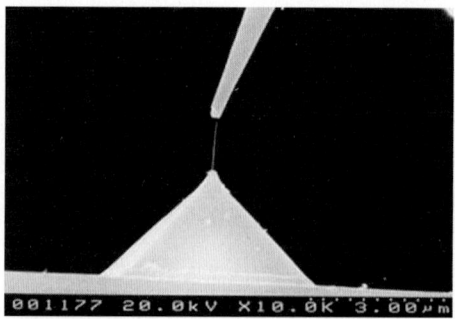

Fig. 17.1. Manipulation of a single nanotube using the nanomanipulator mounted in SEM

microscopy [5]. Very recently, the position and angle of CNT with respect to the probe is well controlled by focused ion beam (FIB) techniques. Most of the applications are limited to noncontact mode AFM including tapping mode, however, contact measurement is necessary in some research areas such as point-contact microscopy where simultaneous measurement of topography and contact current is conducted. Successful atomic scale measurement of mica surface is reported using contact mode CNT–AFM where short tip length is claimed to be essential [6, 7].

On the other hand, it is rather difficult to fabricate CNT-STM probe because the attached surface is not plane but curved. In addition, the contact area should be conductive enough to detect current. In UHV experiments, the CNT tip should be cleaned by heating so that no impurities can fall into the surface [8, 9]. The coating over the CNT-STM probe is effective when making use of the high aspect ratio of CNT [10].

Thus, this method is straightforward and has merit in choosing a CNT to be attached. However, it is time-consuming and is susceptive of contamination due to beam irradiation.

Direct Growth Method

This method is suitable to mass production of CNT probes. Chemical vapor deposition (CVD) is commonly used, and even single-walled carbon nanotube

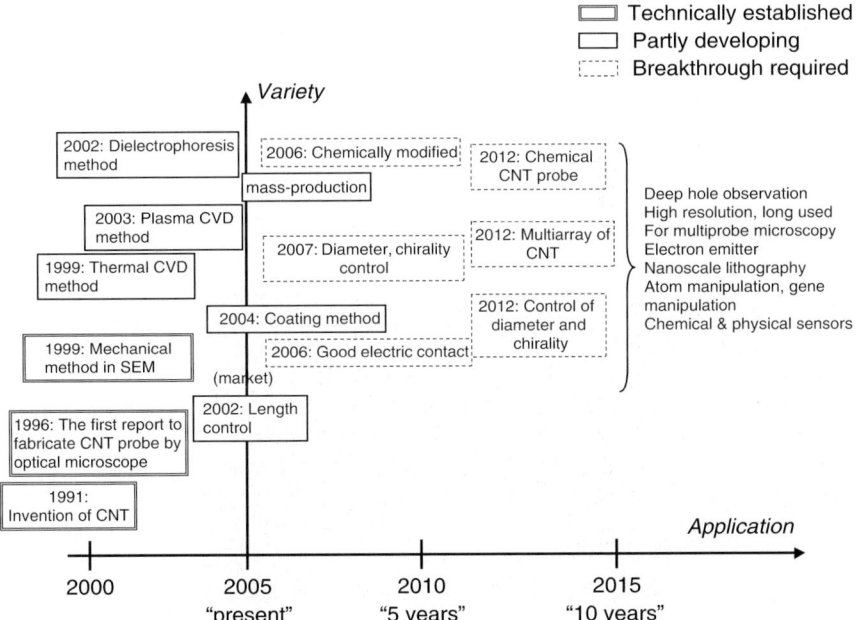

Fig. 17.2. Carbon nanotube SPM probes

(SWNT) can be fabricated. In order to grow CNTs only at the apex, several methods to pinpoint catalysis have been reported [11,12]. In plasma CVD, the growth parameters should be adjusted to balance between etching and deposition [13]. Since the catalysis determines the characteristic of CNTs such as diameter and chirality, this method looks very promising in future. Moreover, multiarray of CNTs will be also fabricated by this method (Fig. 17.2).

Dielectrophoresis

When alternative electric field (MHz) is applied between asymmetric electrodes (for example, metal probe, and counter plane electrode) immersed in CNT solution (water, alcohol, dichloroethane, etc.), CNT is polarized and move to attach the probe (dielectrophoresis) [14–16]. The method is very simple, taking a few seconds, but the SEM observation is necessary to see if CNT is attached or not, and to glue if CNT is. Before observation, heating is effective to eliminate contaminations.

17.3 Roadmap

The CNT-probes come into wide use in SPM before too long. They are also useful as sensors and probes in a variety of scientific and industrial fields including the medical field (Fig. 17.2).

References

1. H. Dai et al., Nature **384**, 147 (1996)
2. S. Akita et al., J. Phys D: Appl. Phys. **32**, 1044 (1999)
3. S. Akita et al., Jpn. J. Appl. Phys. **41**, 4887 (2002)
4. A. Okazaki et al., Jpn. J. Appl. Phys. **41**, 4973 (2002)
5. N. Yoshida et al., Physica B **323**, 149–150 (2002)
6. M. Ishikawa et al., Appl. Surf. Sci. **188**, 456 (2002)
7. M. Ishikawa et al., Jpn. J. Appl. Phys. **41**, 4908 (2002)
8. T. Shimizu et al., Surface Science **486**, L455 (2002)
9. W. Mizutani et al., Jpn. J. Appl. Phys. **40**, 4328 (2001)
10. T. Ikuno et al., Jpn. J. Appl. Phys. **43**, L644 (2004)
11. C.L. Cheng et al., PNAS **97**, 3809 (2000)
12. F.M. Pan et al., J. Vac. Sci. Technol. **B22**, 90 (2004)
13. Yoshimura et al., Jpn. J. Appl. Phys. **42**(7B), 4841 (2003)
14. Japan Patent: 3557589
15. C. Maeda et al., Jpn. J. Appl. Phys. **41**, 2615 (2002)
16. J. Tang et al., Adv. Mater. **15**, 1352 (2003)

Characterization of Semiconducting Materials

Shuji Hasegawa and Masahiko Tomitori

Since scanning probe microscopy (SPM) enables characterizations of surface structures, dynamical processes, and electronic states of semiconductor crystals in atomic scale, SPM is now widely used not only for academic research but also for applications to device fabrication. However, SPM has some weak points as a characterization tool for semiconductors. First, SPM is quite surface-sensitive, which means, reversely, that it is difficult to obtain the information of interior of semiconductor crystals. Even several atomic layers below the surface are hardly detected by SPM in many cases. Second, since the tips/cantilevers are scanned mechanically over the sample surface, time resolution in imaging is insufficient in some cases. Third, although SPM has very high spatial resolution, it is not suited to analysis of large areas such as the whole area of a wafer. Fourth, in spite of the atomic resolution, SPM basically cannot identify the species of individual atoms. Despite of these faults, however, SPM is widely utilized for semiconductor characterization in various ways as shown in Fig. 18.1. This is wholly owing to the atomic resolution of SPM and some new methods developed to partially avoid the faults mentioned earlier. Further improvements in mechanical parts as well as electrical aspects of SPM will be possible and necessary for specific purposes as described later.

18.1 Characterization of Semiconductor Surfaces

It is now routinely possible by SPM to directly image the surface reconstructions, domain structures, atomic steps, surface roughness, atomic vacancies, and adsorbed atoms and molecules, etc. at atomic resolution. And, "in vivo SPM" is also developed, in which dynamical changes of atomic structures during crystal growths, nanostructures formation, and chemical reactions are in situ observed [1]. By using scanning tunneling spectroscopy (STS) and conductance imaging method (so-called dI/dV imaging), furthermore, analysis of electronic structures is also routinely possible with atomic spatial resolution. The SPM has been recently utilized to directly image electric current paths in

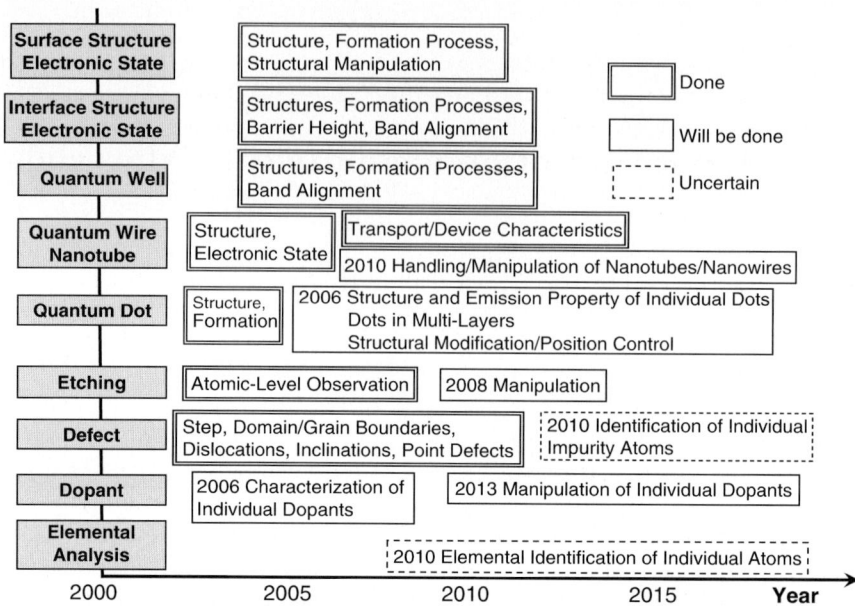

Fig. 18.1. Present status and future prospect about semiconductor characterization by SPM

mesoscopic scales [2] and electron wavefunction [3]. Thus, the SPM can now satisfy almost all requirements of structure analyzes of static and stationary states. However, for lack of time resolution in SPM observations, the SPM does not suite for dynamical analyzes of temporally fluctuating structures and electronic states; the images should be interpreted with aid of theory and assumptions in some cases [4]. To overcome the limit of time resolution in SPM imaging, however, a method is developed in which structural fluctuations are detected in real time as variations in tunneling current by fixing the tip at a specified point, without scanning for imaging [5]. Another method, so-called "atom tracking method," is also developed to trace the diffusion of a targeted atom on a surface at real time, in which the scanning area is limited only near the targeted atom [6]. Both methods, however, can reveal the dynamical phenomena only partially in limited ways and sometimes irritatingly beside the point. On the other hand, high-speed SPM is also developed in which a frame of image is taken on the order of ms, the details of which are described in Chap. 14.

In addition to the imaging of atomic and electronic structures, the SPM is utilized to measure the electrical characteristics of semiconductors. Some examples have been demonstrated; electrical conductivity measurements by four-tip scanning tunneling microscope (STM) (see Chap. 12), by point contact method [7], and by scanning potentiometry by a single-tip STM [8], band bending measurement by using photovoltage phenomenon with light shining during STM observation [9], and so on. These methods can reveal the electrical properties of semiconductor surfaces with high spatial resolution, but we should say that the results are not yet fully utilized for improvements in performance of semiconductor devices as well as for exploring new physics in nanometer scale. In order to break the frontiers of nanoscience, we need to find suitable sample structures as well as to improve the SPM capabilities.

18.2 Characterization of Semiconductor Interfaces

Although the surface phenomena are very important for semiconductor processes such as thin crystal growths and etching, many of the functions of semiconductor devices come from the interface at, e.g., heterojunctions. Therefore, it is quite important to analyze the structures and properties of such buried interfaces. Although, unfortunately, the SPM does not suit well for this purpose, some trials are made. By cleaving semiconductor crystals having quantum well structures and superlattice structures, the SPM is employed to observe the cleaved surface and analyze the band offsets at the heterojunctions [10]. But when the crystal is cleaved and the heterojunctions are exposed to the surface, the band bending can change in some cases. So we need some complementary measurements by other techniques for reliable analysis.

One of the most important features for the semiconductor heterojunctions may be roughness at the interface. The device properties depend on whether the interface is atomically abrupt and smooth. But the SPM is not able to analyze the buried interface roughness at atomic scale in the in-plane direction. Although one of the SPM-derived techniques, ballistic electron emission microscope (BEEM), can reveal the spatial distribution of Schottky barrier at metal–semiconductor interfaces, the interpretation of data is not straightforward in usual cases. Since imaging and analysis of buried interfaces are one of the challenges for the future SPM technology, we may need to combine the SPM with some other techniques such as tomography and magnetic resonance imaging.

18.3 Characterization and Manipulation of Semiconductor Nanostructures

When the size of semiconductor devices is reduced to be on the order of nanometers, comparable to the Fermi wavelength of conduction electrons there, quantum phenomena appear due to the confinement and correlation effects of electrons. We can expect novel functions and properties from such nanostructured semiconductors, and high-speed and low-energy-dissipation devices can be fabricated. For the research along this direction, we need to measure the electrical, magnetic, and optical properties of individual nanostructures, not the averages of assemblies of the nanostructures. For example, it is strongly required to measure band alignment of individual quantum wells, conductivity, and electronic state of individual quantum wires, quantized energy levels and light-emission property and magnetization of individual quantum dots, and so on. Some of them are already done by the SPM techniques. These quantum structures are so small that the SPM can wholly probe the properties. Light emission spectra from individual quantum dots, for example, are directly measured by STM-induced photon emission spectroscopy (see Chap. 8) and near-field optical microscope (NSOM) (see Chap. 4). It has been revealed that the emission property is actually affected by the dot size/shape and interface conditions between the substrate. Atomic structure, chirality, electronic states, and electrical conductivity of individual carbon nanotubes are measured by STM/STS and by two-terminal method using multitip STM [11, 12].

In addition to characterizations of structures and properties of semiconductor nanostructures, the SPM is utilized to manipulate and control the formation processes of nanostructures and also to handle them. Stimulation by tunneling current from an STM tip is utilized to initiate the nucleation of quantum dots at specified positions [13], and also to initiate a chain reaction of polymerization [14].

While, as mentioned so far, great progress has been made in the uses of SPM techniques for characterization and manipulations of semiconductor nanostructures, there remains a lot of issues to be solved. For example, we yet cannot separate semiconducting carbon nanotubes from metallic tubes. We cannot position individual carbon nanotubes at specified positions, either. The SPM may not be suitable for integration of nanostructures as well as mass production of nanostructures. The SPM techniques will be used only for producing a kind of mold which will be used afterward for integration and mass production of semiconductor nanostructures. Thus, it will be important not only to improve the SPM techniques, but also to find the structures and fabrication processes in which the SPM can show the merits.

18.4 Characterization of Defects in Semiconductors

There are a variety of defects on semiconductor surfaces, such as steps, grain/domain boundaries, point defects like vacancies and adatoms, penetrating dislocations, and so on, which are easily observed by SPM. These defects play important roles in oxidation, etching, chemical reactions, and crystal growths. Furthermore, they affect life time and scattering of carriers near the surface, which in turn determine the transport and light emission properties. Although we can know the position, distribution, and density of defects from SPM observations, it is quite rare to demonstrate the influence of defects on the properties by SPM. The SPM must be useful to analyze how much the individual defects shorten the carrier life time and how much electrical resistance the individual defects produce. Four-tip STM has been used to measure the resistance produced by a single monatomic step on an Si crystal surface [15]. Such applications of SPM to measure the properties should be more explored. It will be necessary for this purpose that SPM measurements should be done with changing the environment conditions such as temperature, magnetic/electric fields, light illumination, current flowing through the sample, and under device operation.

The most important defect in semiconductors is impurity dopants. Individual dopant atoms at subsurface region are imaged by STM as standing waves around them [16]. But it is generally impossible to image the dopant atoms in semiconductor crystals. We yet cannot identify the atomic species of individual dopants, either. For this purpose, we need some improvements of STS techniques combined with section imaging techniques.

18.5 Characterization of Semiconductor Processes

Surface reactions such as crystal growths, oxidation, etching, metal film condensation, and silicide formation play main roles in the processes of semiconductor device fabrication. The SPM is used to observe the atomistic phenomena in these processes, and the results are utilized to optimize the conditions in the processes and subsequently improve the device performance. Local chemical reactions such as oxidation and etching can be induced by SPM probes. But the ultimate control of the process like doping of individual dopant atoms in semiconductor crystals with controlled manners is not yet done. The SPM may have potentiality for such atomistic controls.

References

1. B. Voigtlander, Surf. Sci. Rep. **43**, 127 (2001)
2. M.A. Topinka et al., Nature **410**, 183 (2001)
3. M. Morgenstern et al., Phys. Rev. Lett. **89**, 136806 (2002)
4. Y. Nakamura et al., Phys. Rev. Lett. **87**, 156102 (2001)
5. K. Hata et al., Phys. Rev. Lett. **86**, 3084 (2001)
6. B.S. Swartzentruber, Phys. Rev. Lett. **76**, 459 (1996)
7. Y. Hasegawa et al., Surf. Sci. **357/358**, 32 (1996)
8. S. Heike et al., Phys. Rev. Lett. **81**, 890 (1998)
9. M. Mcellistrem et al., Phys. Rev. Lett. **70**, 2471 (1993)
10. R.M. Feenstra, et al., Phys. Rev. Lett. **72**, 2749 (1994)
11. H. Kim et al., Phys. Rev. Lett. **90**, 216107 (2003)
12. S. Frank et al., Science **280**, 1744 (1998)
13. M. Shibata et al., J. Electron Microsc. **49**, 217 (2000)
14. Y. Okawa, M. Aono, Nature **409**, 683 (2001)
15. I. Matsuda et al., Phys. Rev. Lett. **93**, 236801 (2004)
16. M.C.M.M. van der Wielen et al., Phys. Rev. Lett. **76**, 1075 (1996)

Evaluation of SPM for LSI Devices

Koji Usuda, Takashi Furukawa and Yasushi Kadota

19.1 LSI Devices and Forecast

19.1.1 Development of Si-LSI Devices

Development of Si-LSI devices has proceeded for almost 30 years according to the scaling law. The basic point of the scaling law is that it is possible to manage process cost, improvement of device performance, and integration while keeping the inner potential of a metal oxide semiconductor field effect transistor (MOSFET) to be constant as the device integration proceeds. Table 19.1 summarizes the specifications for an Si-MOSFET, which is a basic component

Table 19.1. Summary of ITRS 2004 update

	2004	2007	2010	2013	2016	2018
Technology node	*hp90*	*hp65*	*hp45*	*hp32*	*hp22*	*hp18*
Physical Gate Length						
(nm)	37	25	18	13	9	7
EOT						
(nm)	1.2^a	0.9^b	0.7^b	0.6^b	0.5^b	0.5^b
Metrology for EOT						
(3σ nm)	0.0048^a	0.0038^a	0.0028^b	0.0024^b	0.002^b	0.002^b
Gate leakage current	4.50	9.30^a	1.90^b	7.70^b	1.90^b	2.40^b
($A\,cm^{-2}$)	$\times 10^2$	$\times 10^2$	$\times 10^3$	$\times 10^3$	$\times 10^4$	$\times 10^4$
Dopant conc.	1.5–2.5	2.5–5.0				
($atoms\,cm^{-3}$)	$\times 10^{18}$	$\times 10^{18}$	NA	NA	NA	NA
Junction depth						
(nm)	20.4	13.8	7.2	10.4	7.2	5.6

for HP: High-performance MPU
[a] Means the manufacturable solutions are known
[b] Means the manufacturable solutions are not known

of Si-LSI devices, extracted from the latest version of international technology roadmap for semiconductor (ITRS) roadmap 2004 update [1]. The roadmap suggests that physical gate length of an MOSFET will be 7 nm and a thickness of a gate dielectric oxide layer will be less than 1 nm at the beginning of the 2020s.

In that case, it is pointed out that the electrical power dissipation of an MOSFET device per unit area will rise to an extraordinarily high order. The development of particularly low power and low-standby performance devices is considered to meet significant issues. Moreover, it is anticipated that the devices might show unexpected electrical behavior such as tunneling effect, making it difficult to control the device operation.

Hence, it is suggested that the development of the device performance based on the scaling law might reach its limitation around the 2020s. It is so-called end of Si-LSI roadmap.

19.1.2 Forecast of Si-LSI Devices

Novel device structures and fabrication processes which do not depend on classical-MOSFET technologies have been successively proposed to continue the prosperity of the conventional Si-LSIs. One of the most significant innovations involves the replacement of the gate dielectric film of MOSFETs. The SiON film was employed instead of conventional SiO_2 film for the 90 nm technology node (TN) generation. Then, high-dielectric (high-k) film, such as HfO_x, HfSiON, HfAlON, and ZrO_y, has been investigated with a view to adoption for the 65 nm TN architecture. Another important innovation involves the use of new materials and new fabrication process technologies. For example, technology of silicon on insulator (SOI) structure and strained-channel has been applied for the latest Si-LSI devices.

Furthermore, many studies concerning a limitation of MOSFET performance are already underway. Performance about a very small MOSFET has been reported whose gate length (L_g) was as short as 6 nm [2]. Additionally, investigation of a ballistic conduction within the channel for the ultimate carrier transport mechanism of MOSFET operation is being followed with keen interest.

In regard to the above-mentioned background, Fig. 19.1 shows a forecast of the performance of Si-MOSFET, which is a basic device for Si-LSI systems, through to 2025. The most significant point is that a newly developed three-dimensional structure utilizing various new technologies is expected to be employed in order to overcome the limitations of conventional planer MOSFETs. Even with such new device structures, the principle of the MOSFET is basically the same as that of the classical MOSFET and the technology trend shown in the ITRS roadmap is expected to be maintained until almost 2020.

On the other hand, other new technologies which do not depend on the MOSFET performance itself have already been employed. For example, so-called multithread and multicore technologies have been realized and

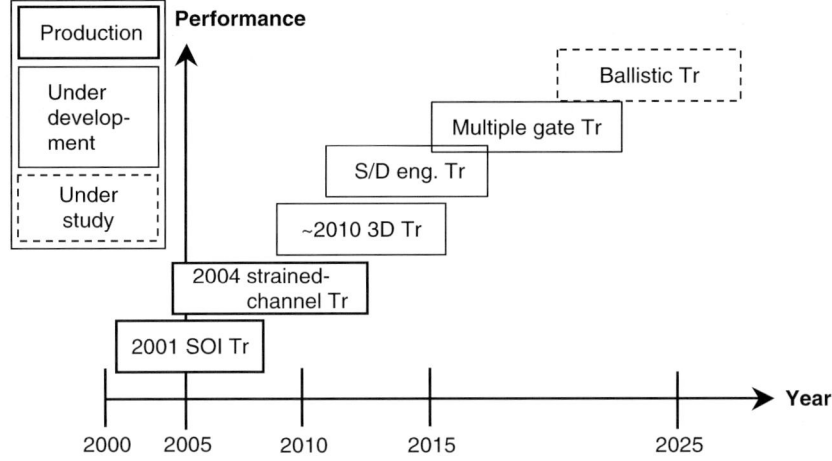

Fig. 19.1. Forecast of Si-MOSFET performance

introduced to the consumer market. Another LSI technique which has been rapidly introduced into LSI systems is the system on chip (SOC) concept, which does not depend on the MOSFET performance.

Consequently, the combination of new device structures enabling higher performance than that of conventional MOSFETs and new functions which will not be realized by the conventional scaling theory can be a technology driver within the framework of the conventional roadmap. In other words, the roadmap can be progressively revised by changing its technology indices.

In addition to the proposals of new device concepts, the establishment of precise control of device processing within nm order or less is necessary for the further development of Si-LSI devices. Therefore, development of high performance devices according to the roadmap is still important and the establishment of measurement and evaluation methods within nm order and/or less is definitely important, for the major indices such as gate length, thickness of gate dielectric films, and shallow pn-junction.

Hence, the present state of evaluation technologies and methods and the latest technology of scanning probe microscope (SPM) are discussed below with the aim of achieving further progress in the development and processing of Si-LSI devices.

19.2 Present Evaluation Technologies of LSI Devices and Latest Trend of SPM Characterization

To achieve the previously mentioned points, it has been predicted that a three-dimensional evaluation at the nanoscale level, an impurities concentration evaluation, and an active characteristic evaluation of the integrated device will be necessary. In other words, the establishment of a measurement technology, and a defect analysis technology, which make use of the characteristics of SPM, for increases in the semiconductor manufacture process at present, are necessary. In this section, technique for LSI devices characterization is summarized and trend on newly developed SPM evaluation methods are discussed.

19.2.1 LSI Device Evaluation Technology

For the accomplishment of a shortening the development period of a new devices, the overall process itself, and the mass production of the devices, various defect analysis technologies have been abundantly developed, and are applicable in the semiconductor manufacture process. However, the acceleration in device development in recent years, which called for ITRS guidelines [1], should make complications and the integration of the device's structure even more remarkable. Once again, we have reached a level where a demand for controlling the gate length and film thickness on nanoscale levels, for the high performance operation of the device, is necessary. Following this, electrical failure of devices increases as the elementary device become small size and the structure become complicate. So, it is indispensable to measure the physical characteristics of the devices and form device's shape with high reliability and precision levels.

With this in mind, although a demand for this type of advanced defect analysis technology continues to rise, it is very difficult to achieve. The demand for typical measurement items in the defect analysis of the processes origin, the present condition of analysis technology, and the future predictions up to 2013 (the year of the 32 nm technology node) are shown in Fig. 19.2. Concentric circles show the needs of a generation and each measurement item, and the radiation-shaped thick-lines show the levels of present marketing devices. It is clear in Fig. 19.2 that measurement and analysis technology are under their needs with most measurement items. This trend has become clear over the past several years. It has become a Red Brick Wall and a Death Valley, so a breakthrough technology based on new general ideas has been demanded for a metrology chapter of ITRS as well.

SPM becomes an indispensable tool for the application of today's actual development of a nanoscale semiconductor device, because of their high space resolution of SPM being the predominant character. So, SPM technology is overlooked, which makes it possible for the development of actual devices.

19.2.2 Latest Trend of SPM Evaluations

Dopant Distribution Measurement Technology

For the development of semiconductor devices, we need to control the three-dimensional career distribution in a semiconductor to get a requested electric character by freely using a complicated and advanced process technology.

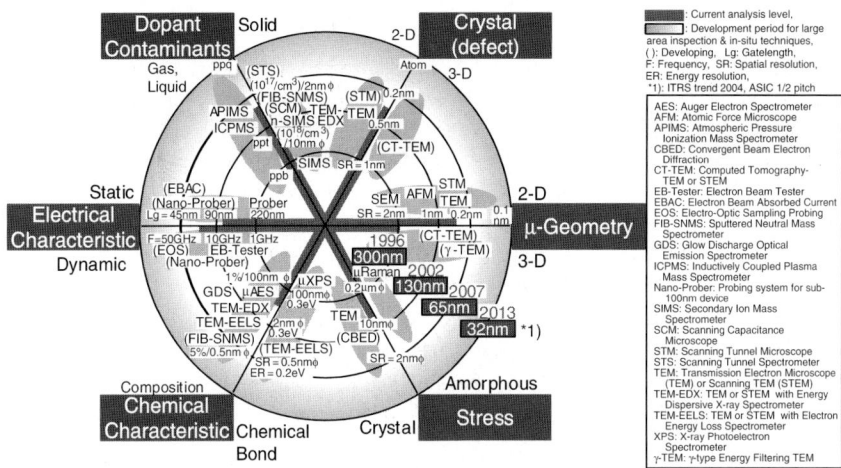

Fig. 19.2. Present analysis technology trend for semiconductor device failure analysis and future prediction (by Dr. Y. Mitsui)

Therefore, this optimization becomes a very important factor in the development stage of a device.

To optimize the dopant distribution in the actual devices, prediction of dopant distribution using a computer simulation is indispensable. However, as the precision of the present simulation is not always high, a comparison between the results of the calculations and the experiments becomes indispensable. Furthermore, as the dopant distribution does not necessarily correspond with the actual career distribution activated by heat-treatment, we need to compare the correlations between the distributions of the dopant and the career.

The dopant and career distribution in the minute area of the semiconductor are experimentally evaluated, and a lot of research is being conducted around the United States [3]. Especially for the scanning capacitance microscope (SCM), a lot of papers have been published on a technique to specially evaluate the two-dimensional career distribution [4]. The space resolution of SCM, however, lessens as the dopant concentration lowers, because it evaluates the spread of a depletion layer in the semiconductor with probe. For example, when impurity concentration in an Si semiconductor is 1×10^{17} cm^{-3}, the space resolution becomes about 0.1 μm, and this is not thought to be enough for an actual device evaluation. The scanning spreading resistance microscope (SSRM) has recently been being researched as a technique to complement this SCM. Like SCM, SSRM is often used to measure two-dimensional dopant profiling in semiconductors, but it does so by quantifying the electrical conductivity or resistivity. In SSRM, an electrically conductive probe is used to measure the sample's local resistively. When the probe is scanned in contact mode over regions with different resistivity, the electrical resistance formed

Table 19.2. Comparison of various methods of measuring dopant distributions

	Resolution (nm)	Detective concentration (atom cm^{-3})	Pseudo-active evaluation	Quantitative analysis
SCM	10–20	$1\times10^{15}\sim$	Possible	△
SSRM	10–20	$1\times10^{15}\sim$	Possible	○
Selective etching	10	$1\times10^{17}\sim$	Impossible	△
STEM/EDX	1–2	$1\times10^{19}\sim$	Possible	○
STM/STS	≤ 10	$1\times10^{17}\sim$	Possible	×
Electron holography	2–5	$1\times10^{18}\sim$	Possible	○

by the probe sample contact will proportionally vary. A major application of SSRM is the measurement of the two-dimensional distribution of electrical carriers inside semiconductor structures [5]. Besides these, though methods such as chemical selective etching depending on the dopant concentration [6], scanning transmission electron microscopy (STEM) with energy dispersive X-ray fluorescence spectrometer (EDX) [7], have been proposed, there are some subjects for solving the difficulties of the quantitative analysis of a dopant concentration, low space resolution, and the low sensitivity for the light element, among others.

Recently, scanning tunneling spectroscopy (STS) owned by STM [8], and electron holography [9] has been used for the rapid development, as a nanoscale career distribution measurement technique. The former is using the idea that an STS signal between the sample and probe depends on the career concentration of the sample and its changes. In this case, the career distribution can be visualized by using a luminance signal translating from a tunneling current, which it can get when a probe is scanned in the fixed height above the sample surface. Although it was developed as a technique to visualize a micro-electromagnetic field, the technique in which the latter's electron holography was used again is being watched even if the career distribution measurement technique inside the transistor has been recently taken. The career distribution in the pn-junction inside the semiconductor is detected with this technique, as a change in the average internal electric potential that an activated dopant atom forms it. Therefore, these methods become applicable to the device doped by a light element, such as Boron, which is difficult to measure using a conventional EDX analysis method for measuring atom detection. Both methods are effective in nanoscale measurements because they are expected to measure in the high space resolution in an atomic order. Various measurement techniques are put together in Table 19.2.

SPM Technology for Large Size Wafers

The recent development of a device has been used, in addition to the planarization of the high step and the usual etching, together with the filling process that is aimed at the improvement in the productivity of the cost reduction using a large size Si wafer (ϕ300 mm). To manage these processes, a new method for a high precision flatness occasion measurement to go into the whole chip's surface, and the processing of nanoscale depth measurement is required. In addition, a conventional method for cutting the wafer to measure the distribution of these parameters has been kept at a distance to avoid the economical loss of expensive large size wafers. Although critical dimension SEM (CD-SEM) plays an important part in surface form observations of a wafer and high precision dimension measurements, researchers are worried about a decline in the measurement precision by a depth of a focus' becoming shallow, caused by a lowering of the electronic voltage of the primary electrons, and the problem of damage to the device caused by the SEM observation.

There has been special attention paid to AFM, which is applicable to large size wafers as a new in-line evaluation device, which solved these subjects. There is proof, which has already given actual results in in-line usage, even in operating machines on the ϕ300 mm line [10].

As a future prediction, application to lithography, which is the application development of a large size SPM, can be expected. Actually, research aimed at this realization has been promoted, and the development of a technology that AFM is used as a technique for measuring the dimension of the pattern in a precision range of 0.5–0.3 nm, is vigorously being conducted [11]. Realization of higher turn around time (TAT) of the AFM measurements compared with that of SEM is considered to be difficult in near future. And it is expected that a local analysis which could not carried out with the SEM due to its spatial resolution or sample structure, will be utilized at the first.

A Method of Microscopically Analyzing of Dielectric Films

Formation of ultra-thin and uniform dielectric gate film is necessary to realize high performance MOSFETs. The electrical properties of the gate film strongly affects on several factors, such as variation of film thickness and quality, tunneling current which is prominent with film thickness of less than 3 nm, and increase of leakage current induced by electrical stress. Although the suppression of leakage current is one of the most important issues concerning the maintaining of the electrical reliability of MOSFETs, the phenomenon arises at very small local area and the specification of the behavior has not been analyzed yet. Thus, for many years there has been a need for an accurate and easy method of detecting the leakage current through thin dielectric film.

Recently, a microscopic analysis for observation of degraded dielectric film has been developed, using a conductive-AFM (C-AFM) method [12]. The method enables the direct observation of nanometer-scale leakage current spots in a stressed gate SiO_2 film in an actual MOS capacitor by detecting very small stress-induced leakage current of less than 0.01 pA order (Fig. 19.3).

This technology is expected to be applied for efficient failure analysis of an appropriate device on an actual LSI wafers. Furthermore, the method would

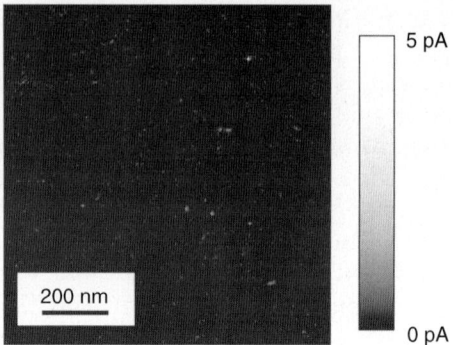

5 pA

200 nm

0 pA

Fig. 19.3. C-AFM image of a leakage current in a stressed SiO_2 film. Leakage current spots (white dots) are clearly seen in the image

contribute to realization of next-generation high-k dielectric film in various ways, such as in clarification of a current leakage phenomenon in terms of a percolation model [13], verification of film crystallization, and observation of the interfacial layer formation between the film and Si crystal.

19.2.3 New Evaluation with Advanced SPM Technology

Multi-Probe Technology for Actual Device Analysis

As mentioned in Sect. 19.1, micromachining technology in the semiconductor manufacture process makes improvements by being a higher density of the element. As shown in first part of Sect. 19.1, various evaluation measurements are indispensable. The electric characteristic evaluation of a special element and/or a transistor is especially very effective in the elucidation of failure movements and the building of a defective model by the physical information, which it could get with other techniques. So far, an LSI tester and a conventional prober have not been used for the electric characteristic evaluation of these devices. However, with an LSI tester, even if a defective cell can be specified, a defective point in each transistor (a bit) unit cannot be specified. With a conventional prober, it is very difficult to measure the electric character of the actual device directly, so a test element group (TEG) with a pad of about 100 μm, is necessary for the contact probe under an optical microscope.

So, multiprobe SPM equipment which measures the electric characteristic of a specific transistor of an actual device was newly developed. Independently controlled mechanical multiprobes in SEM are a type of these equipments. Although the main technology of the multiprobe SPM consists of several probes which can be controlled independently and an SEM, another SPM which utilizes its multiprobe as substitute for the SEM is also proposed [14]. For the first time in the world, the actual characteristic measurement of the specific semiconductor element (single bit) that forms in the sub-micron area

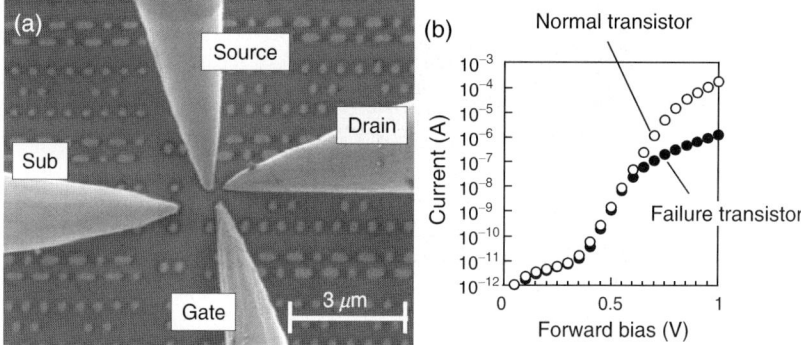

Fig. 19.4. (a) SEM image during measurement of transistor in static random access memory (SRAM) by nanoprober. (b) Characteristic evaluation result of source electrode of transistor for forward bias. Compared with an electrode of a normal MOS transistor, the electrode of the failure transistor shows higher resistance. It is understood that the origin of the failure in the transistor is the high resistance of the contact (by Dr. F. Yano)

on an actual device was found by this multiprobing equipment. An example of a nanoprober measurement [15], which is one of the mechanical multiprobing equipments with SEM, is shown in Fig. 19.4.

An operator of a nanoprober operates the probes in sub-micron precision to a special point in samples under the SEM observation of the sample and the probes, and an electric character is measured with it. The development that precisely controls probes and stages, high resolution field emission SEM (FE-SEM), high throughput using load-lock chambers for sample and probe exchange, a computer-aided design (CAD) navigation system, and easy operation, are based on the new general idea that important element technology is necessary for the achievement of a nanoprober. As mentioned in this section, the development of multiprobing equipment is expected to proceed along with the development of LSI devices. Therefore, the development of both of these element technologies and the ones that are combined with those technologies, becomes indispensable. In the future, measurements under actual movements and by using multiprobing equipment with more than six probes (6-probes are currently used), and the realization of a compound measurement technology with place modification by focused ion beam (FIB) equipment, among others, are expected.

19.3 Roadmap

Recently, research and development of new devices beyond Si-LSIs has been widely performed. Figure 19.5 shows a forecast of development for transistor devices with Si-based ones. One of the features of the forecast is that the

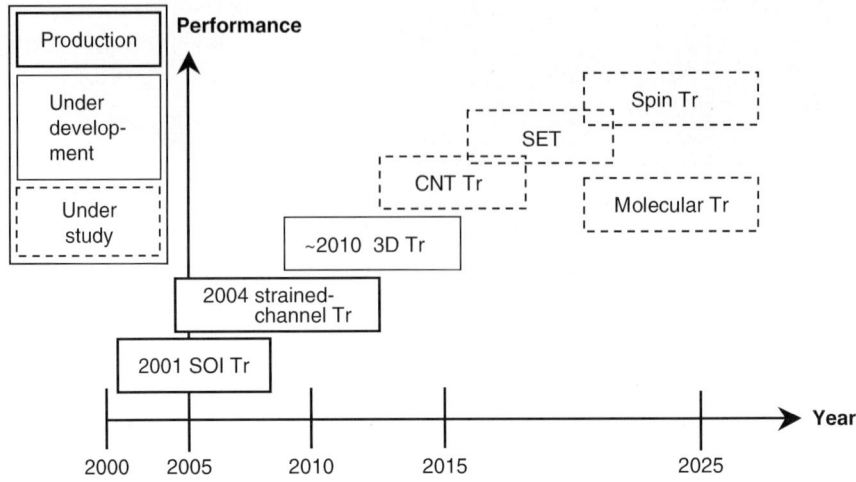

Fig. 19.5. Forecast of transistor development for future LSI devices

research and development is focused on a realization of multi and new functions which cannot be expected to be realized with Si devices.

On the other hand, to maximize a balance between fabrication costs and the new functions, the new devices are not easy to completely replace the Si devices and combination of the new and previous features might be a realistic solution in the development of future devices. Hence, SPM evaluation is expected to be utilized for both devices, as it is an original technology which cannot be replaced by other evaluation methods.

Figure 19.6 shows a forecast of the SPM evaluation methods for future Si-LSI devices. As shown in the ITRS [1], specifications of measurement and analysis for the LSI fabrication processes should become severe with the progress of the technology node of Si-LSI devices, and whether the SPM technology can keep abreast of the LSI development generation or not might become an issue in the near future.

For example, diameter of Si wafer is expected to be 450 mm around 2010 and 600 mm around 2020. Accordingly, newly developed SPM technologies such as fast scan mode and/or multiprobe operation seem to be indispensable.

Moreover, in view of the requirements of LSI development and manufacture costs, a newly developed in-line SPM measurement methods might be required. On the other hand, three-dimensional measurements in nm order are expected to have an important role as the specification of lithography becomes narrower and more complex, and a device structure becomes smaller.

Eventually, SPM be utilized for other would issues such as structural repair of damaged devices using atom and/or molecular probing technology instead

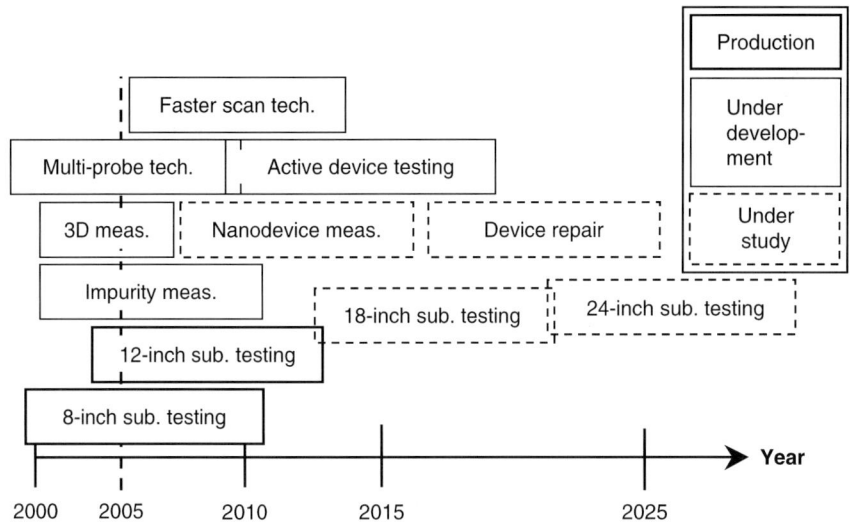

Fig. 19.6. Forecast of the SPM evaluation methods for future Si-LSI devices

of current simple measurement of device length. In any event, new SPM evaluation methods capable of high-speed and wide-area operation are expected to be realized with nm order spatial resolution.

References

1. International Technology Roadmap for Semiconductors (2004 update) http://www.itrs.net/Common/2004Update/2004Update.htm. Cited 10 Jan 2005
2. B. Doris, M. Ieong, T. Kanarsky, Y. Zhang, R. A. Roy, O. Dokumaci, Z. Ren, F.-F. Jamin, L. Shi, W. Natzle, H.-J. Huang, J. Mezzapelle, A. Mocuta, S. Womack, M. Gribelyuk, E.C. Jones, R.J. Miller, H.-S P. Wong, and W. Haensh, in *IEDM proceedings*, (2002) p. 267
3. For example, P. De Wolf, R. Stephenson, T. Trenkler, T. Clarysse, T. Hantschel, and W. Vandervorst, J. Vac. Sci. Technol. **B18**, 361 (2000)
4. For example, V.V. Zavyalov, J.S. McMurray, S.D. Stirling, C.C. Williams, and H. Smith, J. Vac. Sci. Technol. **B18**, 549 (2000)
5. C. Shafai, D.J. Thomson, M. Simard-Normandin, G. Mattiussi, and P.J. Scanlon, Appl. Phys. Lett. **64**, 342 (1994)
6. T.-S. Back, J.-M. Yang, T.-S. Park, H.-J. Kim, S.-Y. Lee, S.-C. Lee, and J.-H. Choi, Jpn. J. Appl. Phys. **39**, 3330 (2000)
7. R. Tsuneta, M. Koguchi, K. Nakamura, and A. Nishida, J. Electron Microsc. **51**, 167 (2002)
8. H. Fukutome, H. Arimoto, S. Hasegawa, and H. Nakashima, J. Vac. Sci. Technol. **B22**, 358 (2004)
9. M.A. Gribelyuk, M.R. McCartney, J. Li, C.S. Murthy, P. Ronsheim, B. Doris, J.S. McMurray, S. Hegde, and D.J. Smith, Phys. Rev. Lett. **89**, 025502 (2002)

10. S. Hosaka, D. Terauchi, H. Sone, T. Morimoto, Y. Kenbo, and H. Koyanagi, Jpn. J. Appl. Phys. **43**, 3572 (2004)
11. S. Gonda, K. Kinoshita, H. Noguchi, T. Kurosawa, H. Koyanagi, K. Murayama, and T. Terasawa, in *Proceedings of SPIE*, ed. by Richard M. Silver, **5752**, 156 (2005)
12. Y. Watanabe, A. Seko, H. Kondo, A. Sakai, S. Zaima, and Y. Yasuda, Jpn. J. Appl. Phys. **43**, L144 (2004)
13. R. Degraeve, G. Groeseneken, R. Bellns, J.L. Ogier, M. Depas, P.J. Roussel, and H.E. Maes, IEEE Trans. Electron. Devices **45**, 904 (1998)
14. http://www.multiprobe.com/ninety_nm.html, and see Application Note AFP2-04 by Multiprobe, Inc. 2003 Rev. 11/04
15. Y. Mitsui, Y. Nara, T. Sawahata, T. Saito, T. Sunaoshi, E. Hazaki, K. Takauchi, F. Yano, H. Yanagita, T. Mizuno, O. Yamada, T. Furukawa, and O. Watanabe, in *Proceedings of LSI Testing Symposium 2005* ed. by Koji Nakamae (Osaka University, Osaka, Japan, 2005), p. 335 (in Japanese)

SPM Characterization of Catalysts

Hiroshi Onishi

20.1 SPM for What?

Our modern civilization is supported by a number of chemical processes for artificial production of useful materials and for removal of toxic compounds from the environment. Heterogeneous catalysts are solid-state devices that provide chemical reactions on demand. Catalytic conversion of chemicals often requires multistep reactions. Reactants are adsorbed on the surface of a catalyst and converted to products via intermediate species. Surface atoms of the catalyst assist reaction steps from one chemical state to another. Different styles and different degrees of assistance are expected on terraces, steps, kinks, and atom vacancies present on the catalyst. Identifying chemical events simultaneously happening over the surface has been ambition of catalyst researchers.

Scanning probes provide potential to achieve the dream. Constant current topography of metals and semiconductors can be observed in laboratories in the world. Insulators are in the observable range of SPM since frequency-modulated AFM (FM-AFM or noncontact AFM) was invented. SPM researchers are now asked: How to reveal the reaction mechanism on a catalyst working in industry? How to provide bland new ideas for technical innovations?

One elegant example was recently demonstrated on a hydrodesulfurization catalyst. Sulfur-containing compounds in petroleum are hydrogenated on MoS_2-based catalysts to release H_2S. This desulfurization process is essential to produce gasoline. The acceptable concentration of sulfur is severely regulated year after year. Besenbacher and co-workers [1] synthesized nanometer-sized crystals of MoS_2 on a gold surface. The atomistic topography observed with their home-built STM was compared with theoretical calculations to conclude metallic electronic states localized at the regions adjacent to the crystal edge. Adsorption and hydrogen-induced reactions of C_4H_4S, a typical sulfur-containing compound in petroleum, were further examined and related to the metallic states. They combined well-known experimental methods (preparing a Au(111) surface, deposition of Mo metal, sulfidation with H$_2$S gas, and controlled dose of C$_4$H$_4$S as well as atomic hydrogen) to bring the catalytically important conclusion.

20.2 Roadmap

Expected developments of SPM technology will open access to research topics as illustrated in Fig. 20.1. Real, industrial catalysts never work in a vacuum. Easy imaging is required in gas and liquid environments. This requirement has already been achieved on metals and semiconductors using STM. An important class of catalyst materials is, however, insulator. FM-AFM provides the true, atomistic resolution when operated in the vacuum, where the Q-factor of the cantilever oscillation exceeds 10^4. The Q-factor is lowered in the viscous environments to reduce the microscope resolution. Latest developments in the low-noise detection of the cantilever motion enable us to achieve high-resolution imaging in an aqueous solution [2]. The author believes that next-generation microscopes compatible in liquids are commercially available within three years.

The ability of scanning probes is beyond the simple topographic imaging. A major class of catalysts contains nanometer-sized transition metal particles interfaced with micrometer-sized metal oxide supports. The density of states are determined on individual gold particles on TiO$_2$ using scanning tunneling spectroscopy and related to the rate of CO oxidation reaction [3]. Another promising method of analysis is scanning Kelvin probe. The charge transfer from a metal particle to the oxide support is thought to control reactions over the metal surface. Macroscopic methods, XPS and AES, have been used to

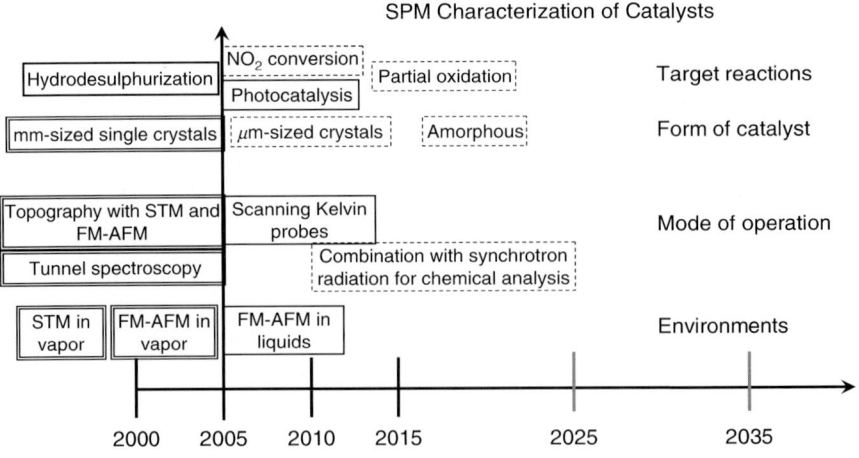

Fig. 20.1. Roadmap of catalyst research using SPMs

detect the chemical shift induced by the charge transfer. Instead, the electron transfer at the interface can be monitored particle-by-particle using a scanning Kelvin probe based on FM-AFM. Actually electron transfer caused by Na adatoms [4] and by Pt particles [5] has been detected.

What are expected for tips? The atomistic resolution is currently achieved only on flat crystalline surfaces of millimeter sizes because of the finite radius of tip apex scanning the sample. This makes a sever limitation to catalyst research aimed at nanometer-sized particles interfaced with micrometer-sized supports. Carbon nanotube tips are promising for imaging granular samples.

Suppose target reactions of research. MoS_2-based catalysts have been beautifully studied as mentioned earlier. Photocatalytic reactions on TiO_2 are also successfully studied. Electrons and holes are excited in the catalysts by ultraviolet light irradiation. The charge carriers migrate to the surface and attach to organic reactants. It is revealed with SPMs where and how the reactants are converted [6]. Extension to dye-sensitized solar cells is possible [7]. Conversion of NO_x to N_2 is of critical importance in reducing air pollution. Catalysts containing CeO_2 work in the exhaust pipe of automobiles. To stabilize the conversion, oxygen is stored in and released from CeO_2 according to the composition of gas emitted from the engine. Oxygen atom vacancies mediating the oxygen transport are ready to be imaged on CeO_2 [8]. Partial oxidation reactions produce various chemicals, textiles, plastics, medicines, etc. Reactants, intermediates, products, and by-products occupy the surface of partial-oxidation catalysts. Imaging such complex surface is definitely requested for future challenge.

References

1. J.V. Lauritsen et al., J. Catal. **224**, 94 (2004)
2. T. Fukuma et al., Appl. Phys. Lett. **86**, 034103 (2005)
3. M. Valden, X. Lai, D.W. Goodman, Science **281**, 1647 (1998)
4. A. Sasahara, H. Uetsuka, H. Onishi, Jpn. J. Appl. Phys. **43**, 4647 (2004)
5. A. Sasahara, C. Pang, H. Onishi, J. Phys. Chem. B **110**, in press (2006)
6. M.A. Henderson et al., J. Am. Chem. Soc. **125**, 14974 (2003)
7. A. Sasahara, C. Pang, H. Onishi, J. Phys. Chem. B **110**, 4751 (2006)
8. C.T. Campbell, C.H.F. Peden, Science **309**, 713 (2005)

21

SPM Characterization of Biomaterials

Atsushi Ikai and Rehana Afrin

21.1 Bioscience

21.1.1 Present Status of Nanoprobetechnology in Bioscience

The use of nanoprobetechnology in the area of bioscience is in image acquisition and measurement of physical properties. In imaging capability of biological samples, nanoprobetechnology is surpassed in resolution by such traditional methods like electron microscopy, X-ray crystallography and nuclear magnetic resonance (NMR) spectroscopy in terms of structural accuracy. In addition, since probe microscopy in general cannot image the atomic arrangements inside the sample molecules, the use of nanoprobetechnology has been rather limited so far in this field. Many results of imaging biological samples by probe microscopy have been reported but most of them are reconfirmation of what have already been resolved by other methods. Very few, among them, contributed to a truly innovative solution of major biological problems by providing new and high resolution images. Nonetheless, since nanoprobetechnology has a unique capability of imaging biological samples under physiologically relevant conditions, active search is underway in many laboratories in the world to exploit nanoprobetechnology in most advantageous way in bioscience.

Nanoprobetechnology has been more aggressively applied to elucidate physical properties of biological samples, because, in this respect, the method is truly new and unique. A nanoprobe can approach, touch, and even penetrate into a small area of the sample and probe the local mechanical, electronic, magnetic, adhesive, chemical, and many other properties in such a way that no other existing methods can even mimic. A problem for the researchers in this respect is more of what kind of physical parameters are most relevant to understand biological functions of particular samples than how to measure them. Biochemistry is based on the chemical reactions inside and outside of the cell and very little attention has been paid to the general material properties of the participating molecules, in particular protein molecules as enzymes and structural elements. Once the three-dimensional structure of an enzyme molecule is determined by X-ray crystallography, the major concern of biochemists is to explain the reaction mechanism of the particular metabolic reaction from the structure of the enzyme and its complex with a substrate analog. The three-dimensional structure of an enzyme is important not only from the point of view of reaction pathways but also from the materials science point of view, in that, the enzyme must be endowed with an optimum local rigidity and flexibility which should be expressed in terms of materials science parameters. Whether such macroscopic parameters for the expressions of rigidity as Young's modulus combined with Poisson's ratio are appropriate for describing the mechanical properties of nanometer sized sample is certainly debatable, but we still need some numerical and quantitative ways for their expression. It is also important to assess the suitability of using such mechanical models developed for macroscopic materials as the Hertz model for the analysis of indentation experiments, since these models are based on the assumption that samples are isotropic, homogeneous, flat, and often nonadhesive [1, 2]. Biological samples often violate these assumptions.

21.1.2 Roadmap 2000

In the previous roadmap 2000, the following problems were given special attentions, therefore, they will be reviewed here:

1. *Use of carbon nanotube probes to improve the resolution of imaging and the development of noncontact AFM for biological use.* Carbon nanotube probes are now commercially available and being used in several laboratories. Resolution is generally improved compared with the results obtained by silicon or silicon nitride probes, but not so high as to satisfy original expectations. Problems of imaging biological samples under physiological conditions may be more in the condition of the sample rather than the quality of the probe.

2. *Harvest of selected portions of biological samples.* By using biochemically functionalized probes, it has become possible to harvest glycoproteins and membrane proteins on the surface of the probe [3, 4]. Chromosomal deoxyribonucleic acid (DNA) harvest is also possible [5]. These results are waiting for the accurate methods of identification of the harvested molecules.

3. *Improvement of immobilization methods of proteins and DNA to the substrate.* Progress in this field has been slow and problematic. A major problem on the application of nanoprobetechnology to bioscience is how to overcome unwanted interactions between the modified probe and the sample surface. Immobilization of biomolecules on solid substrate must avoid the long standing problem of nonspecific adhesion. Proteins and DNA are designed to have specific interactions and reduce the nonspecific interactions as much as possible, but in reality, materials used for substrate and

probes are foreign to them increasing the probability of nonspecific adhesions. Use of protective layers made of anchored polymer cushion is one possibility to safely immobilize biomolecules.

4. *Measurement of cell–cell interaction force.* This has been actively pursued in several laboratories using atomic force microscopy. Interaction forces at the single molecular level is generally in several tens of piconewtons [6].

5 *Development of bionanomanipulators as combinations instruments of AFM with other types of instruments.* It is now common to combine the use of AFM with other types of microscopy and manipulating apparatuses.

In Roadmap 2000, development of high speed AFM was not mentioned but such development has been achieved to a significant level by Ando et al. [7]. His instrument can now image one frame with 80 ms which is fast enough to follow the movement of single molecules of muscle proteins. Using this new AFM many biologically important problems are expected to be solved in the near future.

The following items were mentioned in Roadmap 2000 but have not been realized up to now:

1. Real time visualization of AFM tip, sample, and substrate conditions to help us understand the relationship with the AFM images and force curves. The basic idea is illustrated in Fig. 21.1.

2. Imaging of fibrous molecules like DNA and denatured proteins in extended states.

3. Uninterrupted transfer of sample from liquid to vacuum cells.

Among earlier mentioned, item (3) has not been considered to be acutely necessary, but for items (1) and (2), appropriate methods have not been developed even though they are much needed. In situ monitoring the events taking place between AFM tip and the biological sample is necessary on the cellular level work but because of the limited availability of construction space around the tip–sample area prevents the use of high magnification visualization so far.

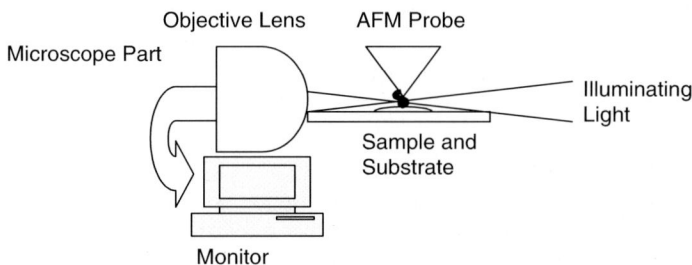

Fig. 21.1. A proposed design of side view microscope attachment to an AFM. An optical microscope with a long focal point objective is used to obtain a side view of the interaction point between the AFM tip and the sample surface so that the operator of the AFM can monitor what is taking place there

21.2 Biotechnology

21.2.1 Nanoprobetechnology in Biotechnology

Nanoprobetechnology using atomic force microscope has been regarded as most promising field for the future development of biological technology with combined effort from physics, chemistry, biology, and engineering disciplines. This trend can be appreciated by the fact that most of the researchers today in the field of AFM application to biology are actually applied physicists with some biophysicists rather than biologists [8]. In the traditional biology, manipulation of biological samples for nonbiological purposes has not been pursued with enthusiasm. Such nonbiological pursuits are more approachable from nonbiological, materials science minded researchers than biologists. It is expected that development in nonbiological field will then be back-applied to biology for a new development of biology.

Let us survey several fields in biotechnology with high expectation for the development of innovative technologies applicable to biologically motivated engineering in the near future. In the previous Roadmap 2000, one focal point was base sequence determination of DNA by using nanoprobetechnology but, so far, the technology has not reached the point to discuss sequencing of DNA based on the visual image of individual bases in a single DNA molecule. In the meantime, biochemical sequencing methods have been further advanced and some of them are reaching 10–100 faster sequencing speeds in the last five years. The visual sequencing based on any kind of microscopic methods has a long way to go to surpass the biochemical methods in speed and reliability. But, because of the possibility that polymerase chain reaction (PCR) amplification step would be unnecessary when direct sequencing is possible, sequencing at the single molecule level using precise molecular visualization technology still remains as an attractive alternative [9].

1. *Base sequence determination of a single DNA chain by molecular manipulation and high speed analysis of genetic traits based on the sequencing.* Estimation of probability of disease expression based on the genetic traits of individuals and the choice of treatments and drugs to be administered to the patients would be done in the future based on the rapid sequencing of the genetic material of individuals. By using methods based on nanoprobetechnology, this can be done rapidly and requiring a very small amount of the genetic samples. For this purpose, base sequence analysis based on the direct visualization of DNA molecules is still actively pursued using nanoprobetechnology methods, especially atomic force microscopy and fluorescence spectroscopy. In the case of amino acid sequence determination of proteins for which no equivalent methods to PCR is available, a rapid visualization for molecular identification should be pursued (10 years). In Fig. 21.2, top and side molecular images of a short peptide are shown. For the amino acid sequencing by direct visual imaging, probe microscopy must be developed to be able to identify each amino acid residues from

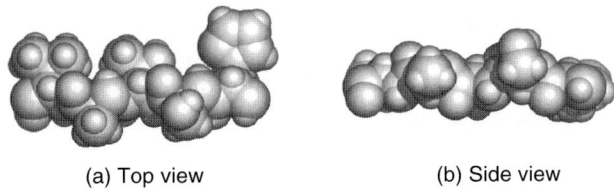

(a) Top view   (b) Side view

Fig. 21.2. Images of a short peptide created by computer graphics. Future probe technology is expected to reach the resolution to read the amino acid sequence of a peptide such as this by directly taking its images

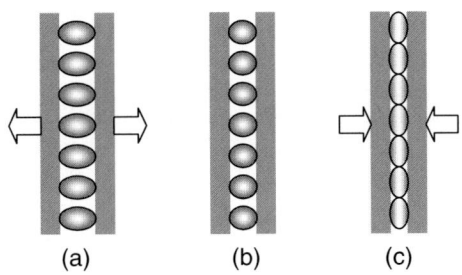

(a)   (b)   (c)

Fig. 21.3. A model of an active biodevice. Enzyme molecules are sandwiched between two parallel plates. (**a**) Center molecules are stretched sideways and not biologically active; (**b**) unstressed and active condition; (**c**) compressed and inactive condition

their shapes and possibly from the contribution of physical properties such as surface charge.

2. *Development of new types of biosensor based on high speed on–off switching operation of proteins and protein-DNA complexes (5 years).* DNA is the information carrier in the cell and from generation to generation, whereas proteins are the devices to express chemical and biochemical functions directly responsible to life processes. Multitudes of protein functions should be utilized as the source of biosensor development but, in reality, only a few of them such as glucose oxidase are used. To bridge this gap, development of biosensors based on their functions is currently actively pursued. In the past, construction of biosensors directly utilized enzyme activity of proteins but in the future, by switching the enzyme activity artificially on and off, more advanced biodevices than simple biosensors will be possible (Fig. 21.3).

3. *Development of self-assembling devices based on the knowledge of biological self-assembly principles (5 years).* In the development of self-assembling devices, a wealth of knowledge can be learned from biological self-assembly systems. Microdevices of artificial design may be possible by mimicking the self-assembling principle of biological structures, such as organs and tissues made from a large number of autonomous cells. New devices can be built by assembling proteins and DNA which will be assembled into

man-made micromachines with highly developed new functions for the benefit of human beings. In every turning point of their development, nanoprobetechnology will play crucial roles in close analogy with the self-assembler suggested by Drexler [10]. Assembling materials may be larger than atoms or molecules and to certain extent they are already half assembled based on the biological self-assembling principles.

4. *Development of identification methods of a very small number of molecules based on the microelectro-mechanical systems (MEMS) technology (5 years).* Analytical instruments of daily use in biochemical laboratories are being made smaller and smaller thanks to the development of microapparatuses produced by the MEMS technology. As the analytical tools are getting smaller, the amount of the samples to be analyzed is also becoming less and less. We need more and more sensitive analytical methods for the identification of the processed samples and, in the end, methods of identification of single biomolecules will be needed. In the present technological development, sensitive fluorescence methods such as confocal microscopy, evanescent microscopy, correlation microscopy are being used for the molecular identification in combination with the specificity of antigen–antibody reactions but they are still far from satisfactory level because of nonspecific reactions and weakness against photobleaching. What is needed is a true single molecule identification method free of contamination errors. In chemistry, single molecule Raman spectroscopy is being developed with rapidly but identification of individual protein molecules by Raman spectroscopy is not an easy task because of the lack of specific signals characteristic to each protein species. To augment the shortcomings of the presently available methodologies, new molecular identification methods based on atomic force microscopy and near field microscopy is waited for.

21.3 Roadmap

21.3.1 Nanoprobetechnology in Biological Field

Let us take a look at the possible advancement of nanoprobetechnology in the biological field in the next five to ten years. First we have to emphasize the special requirement of nanoprobetechnology when it is applied to bioscience. Despite the rapid progress of the nanoprobetechnology in other fields, its application to biological field has been slow. Major reasons for the limited use of nanoprobetechnology in biology may be traced to the following problems special to this field:

1. Nanoprobetechnology must be used in combination with other specialized methods for functional analysis. In biology, function is the most interesting property of all, and imaging structures and measurement of physical properties are all secondary to functions. Therefore, all the measurements should be accompanied by the relevant functional significance.

2. Many studies so far reported emphasized innovative imaging technology of probe microscopy but the results have been mostly confirmation of the facts already studied by other imaging methods. In the future, probe microscopy must be used to make new findings with biological significance. It must be always emphasized that the advantage of probe microscopy is in the ability to touch and manipulate samples and new findings are most likely to come in the combination of imaging and manipulation of the sample.

3. To proceed with the new way of imaging and manipulation of soft biological samples, we need computer simulations as much as possible. The physical interaction between the tip and the soft sample is difficult to understand and the images resulting from such interactions are difficult to be interpreted. There could be pitfalls and the possibility of misinterpretation when we first encounter the images of unknown structures. Measurement of physical properties of soft biological samples is also dotted with difficulties and their interpretation should be attempted with the use of sophisticated computer simulations.

Based on the above inspections, several possible new application of probe technology can be given as summarized later:

1. *Analysis of structure and function of membrane proteins.* Membrane proteins are regarded as good candidates for the development of medicines and drugs that can control the cellular functions. Expectation is high for probe microscopy to contribute to the development of new technologies that would help developers of medicines and drugs in pharmaceutical companies. In science, elucidation of the molecular basis of taste and/or olfactory senses will be rapidly made and application of such discoveries will be found in the development of new drugs. Pheromone receptors are another interesting field of research with potential application in animal husbandry and stockbreeding. Nanoprobetechnology should find a wide application in these fields. In Fig. 21.4, an idealized scheme of membrane harvest for further biochemical analysis is given.

2. *Functional imaging of intracellular structures.* Structural analysis and special distribution of biologically functional molecules inside of the cell may be a good target for nanoprobetechnology. For this purpose, probes must be chemically and biochemically functionalized. One example of such studies is the probe with the capability of ultrasonic tomography which will visualize noninvasively the interior of the cell with ~10 nm spatial resolution with real time imaging. With certain allowable invasiveness, the probe may be inserted into the soft material and visualize the interior of the samples.

3. *Microscopic manipulation and surgery of the cell and chromosomes.* In the case of regenerative biology and medicine where new tissues and organs are produced by changing the cellular properties, techniques to manipulate

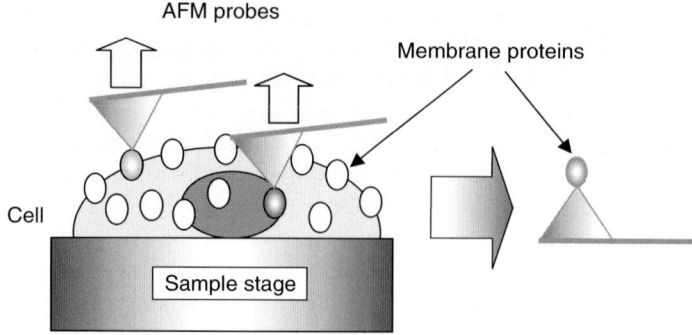

Fig. 21.4. Schematic view of harvesting protein molecules from the live cell membrane. The AFM probe is biochemically functionalized with either bifunctional covalent cross-linkers or specific ligand molecules such as lectins with reactivity to membrane proteins

and multiply cells, and chromosomes with precision will be cultivated by nanoprobetechnology.

4. *High resolution imaging of biologically functional molecules at the single molecular level and evaluation of their physical properties.* In biochemical field, there has not been appropriate methods to evaluate physical properties at the single molecular level, nanoprobetechnology will contribute to provide such properties for proteins, DNA, polysaccharides, lipids, and molecular complexes. The results of such measurement will form the basis of the production of innovative materials of superb properties.

The research areas mentioned earlier are only some examples and the cooperation between the specialists in nanoprobetechnology and those in bioscience will open a fruitful and beneficial research fields in the future. Simple visualization of biomolecules by nanoprobetechnology is in itself an important contribution to biology but the true capability of probe technology is in its power to actually touch and manipulate the sample. No other microscopic technology has this capability. To take advantage of this characteristics of nanoprobetechnology, we have to understand the shortcomings of the present day nanoprobetechnology. First, the present day probe microscope may only either push or pull the soft biological samples. In order to accomplish manipulation of macromolecular complexes and give surgical operations to the living cells and chromosomes, a larger freedom of probe movement and more flexibility in probe modification must be accomplished. Considering these points, the following roadmap is drawn.

Imaging by AFM

1. *Improvement of resolution of AFM (5 years).* Resolution of AFM image for biological samples is approximately 1 nm which far behind that for

inorganic samples such as the silicon surface. By improving the mechanical and electronic features of AFM as well as sample preparation processes, this limit must be improved up to 0.1–0.2 nm so that true atomic resolution can be attained.

2. *High speed imaging by AFM (5 years).* Imaging of biological reactions in real time will be attained by an AFM with time resolution of 1 ms per frame of at least 128×128 pixels, since most of dynamic movement of proteins is considered to be in this time range. This is about 50 times faster than the fastest AFM of present day.

3. *Simulation of biosamples based on their physical properties (3 years).* Imaging by AFM is based on the mechanical, tribological, and electrical interactions between the AFM probe and the sample surface. It is therefore important to understand the relationship between the images taken by the AFM and the physical properties of the sample. The image obtained by the AFM may not correspond to the surface topography of the sample due to the reversible or irreversible deformation of the sample under the interaction with the probe.

4. *Noninvasive imaging technology of AFM on biological samples (5 years).* As stated earlier, physical interaction between the sample and the probe may deform the former. It is, therefore, important to develop ways to image biosamples with minimal deformation or invasion.

5. *Observation of interior of living cells by high performance modified tips (5 years).* AFM usually images only the surface of the sample but it will be possible to image internal structure of biological samples by AFM. One way of achieving this goal should come from new technologies of sample preparation such as gentle and precise methods of opening up a cell and expose its interior. Another possibility is to develop methods to insert an AFM probe into the interior or a live cell and image its inside.

Manipulation by AFM

1. *Development of innovative AFM for the precision measurements of physical properties of biological samples (3 years).* In biology, the importance of physical properties of macromolecules fore their expression of biological functions is not fully appreciated. But this fundamental question must be explored further because it would answer such questions as "Why proteins must have well defined 3D structures for their functions?."

2. *Improvement of modification procedure of the AFM probes for the manipulation of biological structures.* The Uniqueness of probe technology comes from its ability to touch and manipulate samples as well as imaging them. This unique property must be exploited to its fullest extent. For sample manipulation beyond simple compression, chemical and biochemical modification of the probe is indispensable technology. We need more versatile ways of probe modification than available today in combination with biochemical and molecular biological methods. Proteins must

be specifically modified for detailed assessment of their physical properties and their dependence on the subtle change in their structure.

3. *Probe modification for further application of nanoprobetechnology using AFM (3 years).* Analysis of biochemical state of a single live cell will be an important addition to the proteomic methodology of present day. AFM will be used to collect functional molecules from the surface and the inside of a living cell for the further analysis of their identity.

21.3.2 Bionanotechnology

Next we will survey the types of nanoprobetechnology to be needed in the next ten years. Nanoprobetechnology is expected to contribute to the development of bionanotechnology in the following fields. Some of them have been treated as necessary in the section of bioscience.

1. *High resolution imaging of biomolecules by Nanoprobetechnology.* Non-contact atomic force microscopy is the most promising technology in the advancement of image resolution among various probe microscopy. In the future, by using this method in liquid, base sequence of DNA and ribonucleic acid (RNA), and amino acid sequence of proteins should be determined at the single molecular level. In other words, only when sequencing of DNA and proteins is accomplished from their images, nanoprobetechnology can be said to have made a significant contribution to bionanotechnology.

2. *Technology to pick up, transfer, and place individual molecules with nanotweezers (5 years).* For assembling higher order structures from self-assembling molecules, artificial help with nanotweezers will be indispensable. Currently, development of nanotweezers is undertaken using carbon

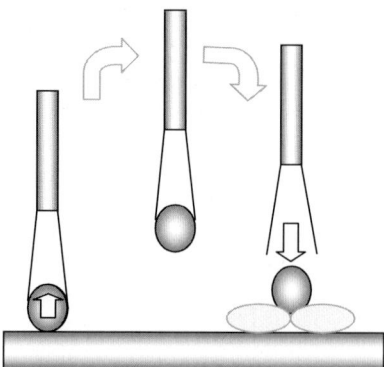

Fig. 21.5. A model of molecular transfer device. From left to right, the device picks up a globular molecule, transfers it to a different site, and deposits on a desired position

nanotubes with electrical switching system to open and close the tweezers. The operating condition is limited to vacuum or in dry air. To develop tweezers which can be operated in liquid environment is essential for biological samples. Therefore, mechanically driven, rather than electrically driven tweezers are desirable. Figure 21.5 gives a schematic image of nanotweezers.

3. *On–off switching technology of protein functions using optical, electrical, and mechanical methods (5 years).* For precisely controlled artificial manipulation of biomolecules and biostructures such as cells or chromosomes, the device technology that can reversibly control the biological functions of protein molecules is needed. Functions of biomacromolecules can be switched by mechanically compressing or extending them [11]. This basic fact must be studied more fully and applied to the development of a new kind of biodevices.

4. *Functional identification of individual biomolecules at the single molecule level (3 years).* This topic has been treated earlier but will be an indispensable technology in the molecular manipulation in the future.

References

1. L.D. Landau, E.M. Lifshitz, (translated by Sykes, J.B. and Reid, W.H.), *Theory of Elasticity* (Course of theoretical physics vol. 7), (Butterworth-Heinemann, London, 2002), pp. 26–31
2. K.L. Johnson, *Contact Mechanics* (Cambridge University Press, Cambridge, 1985) Chap. 4, pp. 84–103
3. R. Afrin, H. Arakawa, T. Osada, A. Ikai, Cell Biochem. Biophys. **39**, 101 (2003)
4. R. Afrin, A. Ikai, FEBS Lett. (in press)
5. X. Xu, A. Ikai, Biochem. Biophys. Res. Commun. **248**, 744 (1996)
6. E. Evans, V. Heinrich, A. Leung, K. Kinoshita, Biophys. J. **88**, 2288 (2005)
7. T. Ando, N. Kodera, E. Takai, D. Maruyama, K. Saito, A. Toda, Proc. Natl. Acad. Sci. USA **98**, 12468 (2001)
8. S.M. Lindsay, in *Scanning Probe Microscopy and Spectroscopy* ed. by D. Bonnell The scanning probe microscopy in biology, (Wiley, New York, 2001) Chap. 9, pp. 289–336
9. J.J. Kasianowicz, E. Brandindagger, D. Brantondagger, D.W. Deamer, Proc. Natl. Acad. Sci. USA **93**, 13770 (1996)
10. E. Drexler, *Engines of Creation: The Coming Era of Nanotechnology* (Anchor Books, 1986, ISBN 0385199732)
11. T. Kodama, H. Ohtani, H. Arawaka, A. Ikai, Appl. Phys. Lett. **86**, 043901 (2005)

SPM Characterization of Organic and Polymeric Materials

Shukichi Tanaka and Ken Nakajima

Here, we describe the roadmap of scanning probe microscope (SPM) characterization of organic and polymeric materials from the point of view of needs. The first discussion will be made separately on both organic and polymeric materials, which will be followed by the anticipation to the future trend of SPM in terms of common demands.

22.1 Characterization of Organic Materials

According to the recent progress of nano-technology, it has not been only a dream to construct the materials and structures as desired by accumulating individual atoms. The pioneer work presented by IBM group in 1992, in which atomic scale characters were created by putting Xe atoms on Ni(110) surface using STM manipulation techniques, has shown the concept is coming to real in near future [1]. This is one of the concepts of nano-technology called as "bottom-up style construction". This new concept has got much attention from the researchers in various fields because of the potentials and promises to bring drastic break-through to the material science and device construction. However, actual material is composed by huge number of atoms, and that it takes tremendous efforts and times to construct the desired structures from individual atoms. In order to overcome the difficulties of bottom-up style construction, there are trials to utilize the various functionalities and flexibilities of organic molecules, especially their self-assembling features. Organic molecules, such as porphyrins and phthalocyanines, have very complex structures composed by several hundreds of atoms, and their physical and chemical properties, such as base structures, chemical reactivity and so on, can be designed as desired by means of chemical synthetic techniques. In that sense, a series of organic molecules can be regarded as the structure, which is constructed by connecting individual atoms as desired. It should be reasonable that the structures and materials constructed by the organic molecules using their self-organization process like LEGO blocks, can be regarded as the structures constructed with "bottom-up style structuring." In order to realize the concepts, accumulation of fundamental knowledge of the physical and chemical properties of molecules are indispensable. In particular, detailed conformation and self-assembling features of individual molecules on the surface must be investigated with sufficient spatial resolution. SPM related techniques, such as STM and AFM, are well-suited for this purpose, because they can flexibly access various kinds of properties while showing excellent spatial resolution. Furthermore, by using SPM probes as a tool to manipulate molecules on the surface, higher complex structures can be evolved.

From the view point of recent achievements and the directions of future progress in the research field, the key topics and trends are roughly summarized in Fig. 22.1. As the progress and improvements of SPM techniques in the past years, visualization of the detailed conformations, inner structures and self-assembling features of organic molecules dispersed on the surface has been achieved, if the experimental technique is limited to STM. As an example, the results of STM observation on the self-organization phenomena of porphyrin-based molecules have shown in Fig. 22.2, whose self-organizing features were controlled using chemical synthetic treatments [2]. It should be noted that the features brought by chemical treatments on individual molecule units can be recognized very clearly. This result implies the important suggestion to chemically control their self-organizing features, and is hardly obtained by means of other experimental techniques, which shows the potentials of STM techniques [3, 4].

On the other hand, the limits and problems of STM technique has been also revealed. Because STM uses a tunneling current to detect the distance

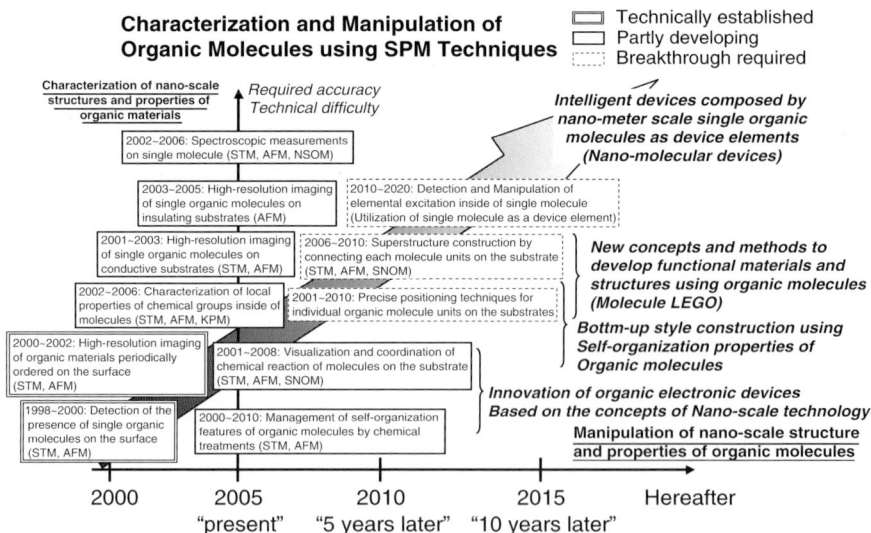

Fig. 22.1. Roadmap concerning the characterization of organic materials/molecules

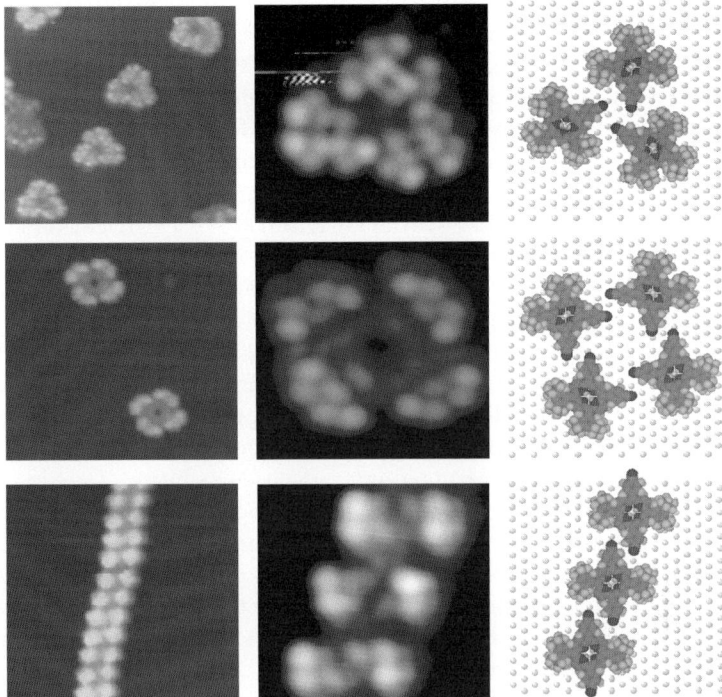

Fig. 22.2. An STM image of porphyrin-based molecules on the surface, whose self-organization features on the substrates are controlled by chemical treatments [2]

between the tip and the surface, the conductivity of the surface is indispensable, and that the properties to be obtained using STM have to be coupled with electrical currents between the STM tip and the surface. This technical limitation is problematic in order to investigate the properties of organic molecules. Because organic molecular structures are held together by covalent bonds, they have no electrical conductivity in many cases. Besides, it is still not clear what kind of properties of molecules are visualized by means of STM, because STM detects the spatial distributions of electronic tunneling probabilities.

In the recent years, non-contact atomic force microscopy operated by frequency modulation mode (FM-NCAFM) has got much attention to complement or replace the STM techniques. Because FM-NCAFM uses a near field atomic force to detect the distance between the AFM probe and the corrugation of the surface, whereby the conductivity of the sample surface is not necessary for measurements. However, a complete understanding of what an FM-NCAFM image means, i.e., detailed characteristics of the near-field interaction between the AFM probe and the sample surface, still seems not an easy task. In addition, only a few reports can be found concerning the observation of individual molecules using FM-NCAFM with sufficient spatial resolution.[1] Indeed, the spatial resolution of FM-NCAFM on organic molecules, which have been reported for porphyrin and phtalocyanin molecules, is, in most cases, much inferior to the one of STM. However, FM-NCAFM and its related techniques are the powerful and promising tools to investigate various kinds of local properties of organic molecules, such as dissipations of mechanical energy from AFM probe, spatial distribution of electrical and chemical potential and so on, without damaging or unfavorable modification. It should be noted that the properties obtained is attributed to the specific part of the FM-NCAFM image of molecules, which is obtained simultaneously at probe scanning. And besides, by comparing the differences and similarities between STM and FM-NCAFM quantitatively, their intrinsic imaging mechanisms and features would be deeply understood. For this purpose, it is strongly desired that the spatial resolution of FM-NCAFM imaging should be improved to the same level as the one of STM imaging[2].

One of the difficulties of FM-NCAFM imaging is the complicated feedback mechanism of the method. In order to realize the reliable feedback control in FM-mode, the electronics and actuators are required to work in very sensitive and stable at the same time, which are not likely compatible. Another factor is that the interactive force between the AFM probe and the surface is very weak and its decay length is rather long compared to the case of STM imaging. Because of that, the AFM probe must approach very close to the surface, resulting in the displacement or vibration of molecules just below the probe. This is crucial because the organic molecules are weakly bound on the substrate by van der Waals interactions in many cases, and they can be easily perturbed by a scanning probe. This weakness of attractive interaction between molecules and substrates is one of the driving forces to realize the self-organization of molecules. Figure 22.3 shows high-resolution FM-NCAFM image obtained by adopting the attractive force introduced by chemical synthetic treatments [5,6]. In the case of Fig. 22.3, the chemical group containing sulfur atoms are introduced into the molecules, which emphasizes attractive interactions to the gold surfaces. It should be noted that the four phenyl-based chemical groups connected to the central porphyrin ring can be clearly recognized in the FM-NCAFM image. The spatial resolution of the FM-NCAFM image is almost the same as the one obtained by STM experiments to exactly the same sample surfaces. This is the first successful example of FM-NCAFM imaging of single organic molecule individually dispersed on the substrates, in which inner structure with sub-molecular scales are finely resolved. From the careful discussion for the results of FM-NCAFM imaging of molecules,

---

[1] See Chap. 3 of this book.

[2] There can be found some preliminary examples of AFM observation for organic molecules, adopting the tunneling current to detect the tip–sample separation like STM.

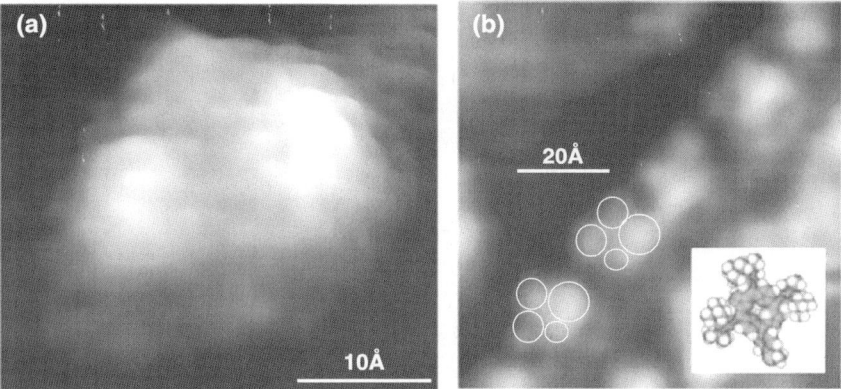

Fig. 22.3. A high-resolution FM-NCAFM image of porphyrin-based molecules on the surface

it is understood that to realize the stable operation, the appropriate control of the mobility of molecules on the substrates, i.e., the appropriate config-uration of the interaction between molecules and substrates, is important. Such a concept is also important to use FM-NCAFM as a fine-manipulator of nano-meter scale molecules on the substrate. For the bottom-up construc-tion using organic molecules, next step is to chemically connect the molecule units on the substrates. For this purpose, the method to investigate chemical properties of molecules have to be established. In detail, the performance of spectrum methods, such as NSOM, KPFM and so on, should be improved to investigate the local features of excitation and relaxation states of individual molecules at their chemical reaction to form or deform the covalent bond with sub-molecular resolutions. By establishing the scheme to control the chemical reaction process with single molecular scales on the substrates, the great flex-ibility in designing materials and structures will be given to the researchers, and that the concepts of "bottom-up style construction" will be very common.

22.2 Characterization of Polymeric Materials

The mainstay of polymer industry is composed of three articles, i.e., textile, rubber and plastic. It is no doubt for these articles to be of great impor-tance among the whole industrial circles. Polymeric materials have a lot of interesting electronic or optical characteristics, while SPM techniques do not contribute to these properties much far beyond other characterization tools at least at this moment. Instead, the advantage of SPM, especially atomic force microscope (AFM), has been properly recognized when polymers as structural materials has to be investigated. Therefore, our emphasis in this section will be placed on the AFM roadmap from the point of view of material scientists as shown in Fig. 22.4.

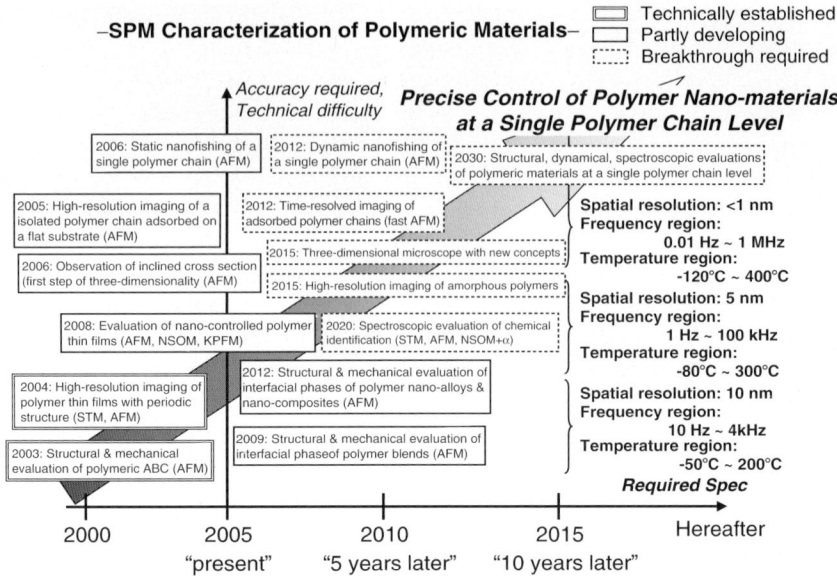

Fig. 22.4. The SPM roadmap for the characterization of polymeric materials (needs side)

Nano-technology is also on the upward trend in polymer industry. In fact, the structures in nm-scale are exceedingly essential in treating polymeric structures and properties, e.g., amorphous polymers, the lamellar thickness of crystalline polymers, micro-phase separated structures of block or graft copolymers, cross-link point of rubbers, the interfacial phases between polymer alloys, blends, and composites (polymer ABC). Polymer nano-composites, where the size of composite materials is in the range of nm-scale, have also drawn much attention in the recent studies. These structures, as well as their molecular-level dynamics, are the control parameters in order to realize polymeric nano-materials [7]. The social demands against these materials at present and in the future can be widely spread; lightweight, high-strength, highly elastic materials, surface functional materials, thermoplastic elastomers, fuel-efficient or wear-resistant tires, separation membranes, medical biomaterials, sensor materials can hardly be realized by homogeneous materials. Thus, polymeric nano-materials, which is essentially heterogeneous materials, have drawn much attention these days and furthermore AFM has been recognized as a very powerful tool to perform structural analyses on such materials.

However, researches to date have mainly focused on structural observations, or at most qualitative characterization of mechanical properties. Quantitative analyses must be a key to expand the marketing share, especially on the interfacial structures and mechanical properties for heterogeneous constituents. Since many polymers are in glassy or fluid states except for some specific conditions, and thus amorphous, any high-resolution imaging is not possible or required in many cases, although such a technology has to be established in the future to increase the AFM ability. In addition, different from other materials, the elastic moduli of polymers are extremely low, resulting easily in the deformation by nN or pN-level forces. One might feel that this is a serious disadvantage, while polymer scientists turns this disadvantage into an advantage with developing characterization methodologies of sample elasticity in quantitative ways. Three items are important in this point; an adequate selection of mathematical models to describe contact mechanics, a precise determination of cantilever spring constant, and a sufficient estimation of probe tip shape. As for the latter two, there have been many different techniques developed in recent years. Especially, it was reported that the tip shape must be treated as hyperboloid instead of simple spherical or conical shapes to perform better quantitative measure [8]. In any case, the most important factor is the former item, i.e., the selection of mechanical models. Several models have been widely used for the purpose such as Herzian or Johnson–Kendall–Roberts (JKR) theories. Furthermore, a guiding principle have been established which models are most appropriate, depending on the degree of load, sample elasticity and the existence of adhesive interactions [9]. Any new design for non-classical model may be necessary in the future, however, which must be preceded by checking the applicability of these classical models at least for a few years.

As easily understood, the AFM images for polymeric materials in many cases do not represent real topographic features due to sample deformation neither in contact mode nor in tapping mode. Even if quantitative analyses on surface elastic properties were successfully made by the above-mentioned methods, the failure in obtaining real topography causes the misleading interpretation because both unreal apparent topographic and mechanical images would contradict to each other. Recently, a method was developed where a sample deformation image was obtained from force–distance curve analysis and it was used to compensate an apparent topographic image into a real topographic image [10]. The method must urgently be applied to the study on nano-alloy and nano-composite systems.

The formulation of tapping-mode operation increasingly becomes important in conjunction with the argument of energy dissipation. As recognized by many researchers, the problem has arisen in the interpretation of phase images because phase shift should occur not only in the case of viscous contribution but also in the existence of adhesive hysteresis [11]. Since polymeric materials have elasticity and viscosity at the same time, in other words, viscoelasticity by nature, the interpretation becomes more complicated. Thus, a certain investigation is really required whether a sample viscosity is properly observed or not. Polymers in general obey time–temperature superposition principle as the resultant effect of viscoelasticity. Soft materials can behave as hard materials by temperature lowering or frequency increase (shortening of characteristic time scale) [12]. It has become possible to purchase a commercially available AFM instrument with superior temperature control, while it is required to develop a novel instrument capable of wide-range frequency sweep to provide complementary information. It is difficult to use cantilever resonant frequency for the purpose. A certain breakthrough is inevitable. Figure 22.4 shows regions of temperature and frequency required in the near future.

It is necessary to realize a single polymer chain experiment for the future prospect. It is a dream in the field of polymer science to characterize and to precisely control a single polymer chain for the purpose of achieving better macroscopic quantities [13]. The acquisition of three-dimensionality in SPM imaging must be realized within ten years. The observation of inclined cross-section can possibly be the first step toward the direction, however, any new concept should be emerged [14].

22.3 Roadmap

The roadmaps of SPM characterization for organic and polymeric materials have been shown, especially from the point of view of material scientists as shown in Figs. 22.1 and 22.4. Another technology to be developed would be the identification of chemical species. Inelastic tunneling spectroscopy (IETS) would prove ultimate goal for the purpose, while the technique cannot be used in realistic materials as they are basically insulators. More practical realization might be near-field scanning optical microscope (NSOM) application. However, fluorescent measurement is only applicable to a small class of material systems. NSOM with capability of Raman scattering detection is a key to promote material science and therefore our effort must be placed on the development of more reliable NSOM Raman systems. Apart from chemical identification, polarized NSOM would also be useful in materials evaluation.

References

1. D.M. Eigler and E.K. Schweizer, Nature **344**, 524 (1990)
2. T. Yokoyama et al., Nature **423**, 620 (2001)
3. J.K. Gimzewski, C. Joachim, Science **283**, 1683 (1999)
4. H. Suzuki et al., Thin Solid Films **393**, 325 (2001)
5. S. Tanaka et al., Thin Solid Films **438–439**, 56 (2003)
6. S. Tanaka et al., Nanotechnology **16**, 107 (2003)
7. T. Nishi, K. Nakajima, *Polymeric Nano-Materials*, (in Japanese), (Kyoritsu, Tokyo, 2005)
8. Y. Sun, B. Akhremitchev, G.C. Walker, Langmuir **20**, 5837 (2004)
9. R. García, R. Pérez, Surf. Sci. Rep. **47**, 197 (2002)
10. H. Nukaga, S. Fujinami, H. Watabe, K. Nakajima, T. Nishi, Jpn. J. Appl. Phys. **44**, 5425 (2005)
11. J. Tamayo, R. García, Appl. Phys. Lett. **71**, 2394 (1997)

12. K. Nakajima, H. Yamaguchi, J.-C. Lee, M. Kageshima, T. Ikehara, T. Nishi, Jpn. J. Appl. Phys. **36**, 3850 (1997)
13. K. Nakajima, Y. Sakai, K. Itoh, T. Nishi, Funct. Mater. (in Japanese) **24**(10), 24 (2004)
14. K. Nakajima, S. Fujinami, H. Nukaga, H. Watabe, H. Kitano, N. Ono, K. Endoh, M. Kaneko, T. Nishi, Kobunshi Ronbunshu (in Japanese) **62**(10), 476 (2005)

Theories of SPM

Masaru Tsukada and Shin-ya Hasegawa

23.1 Present Status of Theories of STM

Theoretical simulation of STM images and STS spectra can be performed by calculating tunneling current between the sample and the tip. Methods for practising this have been well developed from accurate first-principles calculations to simplified tight binding calculations. In a very simple method, the local density of states (LDOS) of sample surfaces is used as the STM image, neglecting the effect of the tip shape or the tip materials. Such an approximation provides a certain idea of the STM image, and not so poor if the delicate tip effect can be ignored and for the case discussing only the STM image, but not the tunnel spectra. An efficient way of theoretical analysis of the STM/STS including the effect of the tip is the method based on the Bardeen's perturbation theory [1, 2]. In the perturbation approach, the parts of the electronic state calculation for the tip and the sample surface can be done only once, making possible efficient calculations for the simulation.

Much more accurate theoretical approaches which are reliable even for the cases with very short distances between the tip and the sample surface, or for the cases under a strong bias field are provided by the method using the scattering waves as the solutions of the Schroedinger equation. The so-called first-principles recursion transfer matrix (RTM) method belongs to this kind [3]. In the RTM method the electronic charge distribution and the potential are calculated self-consistently with the scattering waves. Therefore this method is an extension of the standard DFT theory for open and non-equilibrium systems. In this method, one can calculate the microscopic current in a systematic uniform way from the tunneling regime to the ballistic regime. Thus, the RTM method is a powerful and quite general method for STM simulation, on the other hand, the numerical calculation tends to be considerably large scale compared with those in the perturbation approach. This is because one must calculate the wave functions for the whole system at every scan geometry. Therefore, one should choose the methods of the calculations according to the purpose and target system of the simulation.

Recent interesting development of STM is the measurement of inelastic tunneling spectra by the scanning tunneling spectroscopy. Namely, tunneling current caused by a vibrational excitation of the adsorbed molecule can be detected as a peak in the second derivative of the tunnel current with respect to the bias voltage. This phenomenon provides the information of how the molecular vibration is excited by the tunneling electron as well as the spatial variation of the excitation probability. The theoretical analyses for such phenomena are now developing using phenomenological models [4].

In this problem, further progress of the theories such as the first-principles type calculations including molecular chemisorption mechanism, and effect of electron–vibration coupling on the electron transport through molecules would be necessary. Some interesting reports have been already published on the light emission by the tunneling electron of STM. Such phenomena also await detailed theoretical analyses from atomistic view points. A theoretical model has been so far proposed for the light emitting mechanism due to the interaction of the tunneling electron with the tip induced plasmon [5]. However more realistic calculations should be performed for various systems and confirmed by the comparison with experiments.

23.2 Present Status of the Theory for AFM

The standard theoretical method for the non-contact AFM (ncAFM) is realized by treating the forced harmonic motion of the point mass connected to the spring representing the cantilever motion under deriving force by the feedback circuit as well as the tip–sample interaction force. Namely, this model is described by the following equation:

$$\frac{\mathrm{d}^2 x}{\mathrm{d}t^2} + \gamma \frac{\mathrm{d}x}{\mathrm{d}t} + \omega^2 \left(x(t) - L \right) = F_\mathrm{D}(t) + F_\mathrm{TS}(t) + F_\mathrm{liq}(t). \tag{23.1}$$

In the above, γ and ω are the friction coefficient and resonant angular frequency of the cantilever, respectively. F_D and F_TS are the deriving force of the cantilever by the feedback system and the tip–sample interaction force, respectively. The last term in the right-hand side, F_liq, is the force exerted to the cantilever by the environmental liquid, if the measurement is performed in a liquid. Otherwise this term can be ignored.

Based on (23.1), the phase shift of the cantilever oscillation and its Q-value related with the energy dissipation of the cantilever motion can be obtained, if the tip–sample interaction force F_TS, either conservative or non-conservative, is provided [6]. Then constructing the 2D maps from these observable quantities, we can make theoretical simulations of experimental images of ncAFM. For the calculations of tip–sample interaction force F_TS, the first structure optimized under the presence of the tip is calculated by the first-principles method like density functional theory (DFT), then F_TS is calculated for this optimized structure. For reducing the computational load, more simple calculations as density functional theory adapted tight binding method (DFTB) can be utilized.

In the harmonic oscillator model of (23.1), it is not clear how the point mass model connected with the spring is related with the vibration of the macroscopic elastic body of the cantilever. In particular, the validity of this model is not assured for the ncAFM system in a liquid, and considered not sufficient unless a certain theoretical basis for the term F_{liq} is given.

From a recent theoretical studies performed in the author's group, a clear theoretical understanding on this problem is obtained and the principle to analyse the cantilever motion in liquid is obtained. In this method cantilever is assumed as a continuum of elastic body, and its vibration motion is first solved. The normal modes of the vibration $\phi_n(\xi), (n = 0, 1, 2, \ldots)$ which are the modes existing even in the absence of the external force can be used as the bases for representing any deformation of the cantilever as

$$x(\xi, t) = \sum_n x_n(t)\phi_n(\xi). \qquad (23.2)$$

After substituting the above expression to the equation of the vibration motion of the elastic body of the cantilever, and projecting out on a particular normal mode n, we obtain an equation of motion of the forced harmonic model which is obeyed by the coefficient $x_n(t)$. In doing this analysis, we found how the external forces to the cantilever are mapped on those in the harmonic oscillator model, F_D, $F_{\text{TS}}(t)$, F_{liq}. This analysis is particularly important for the ncAFM in liquid as mentioned later.

23.3 Development of SPM Simulator

During the period 2004–2010, Japan Science and Technology Agency (JST) decided to support the research group organized by the author for the development of the theoretical simulator of scanning probe microscopy, e.g., STM, AFM, KFM (Kelivin Force Microscopy). The members are S. Watanabe (University of Tokyo), A. Ikai (Tokyo Institute of Technology), N. Sasaki (Seikei University), N. Watanabe (Mizuho Information Research Lab.), T. Sasaki (Apriori), and M. Tsukada (Waseda University). In the following, a brief introduction of the JST project and its activities, and a future prospect will be described.

So far, theoretical simulations of SPM have been studied in the research level, in various groups including the authors one. As a typical example in which theoretical simulation contributed illuminating information which leads to solve puzzles of experiments, the STM and ncAFM images of Si$(111)\sqrt{3} \times \sqrt{3}$-Ag surface will be noted. The comparison of the experimental images and their theoretical simulations are shown in Figs. 23.1 and 23.2 for STM and ncAFM at room temperature, respectively. It is remarkable that STM

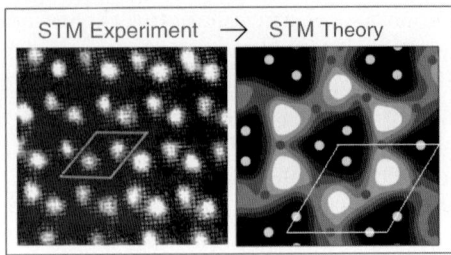

Fig. 23.1. STM image of Si(111)$\sqrt{3} \times \sqrt{3}$-Ag surface

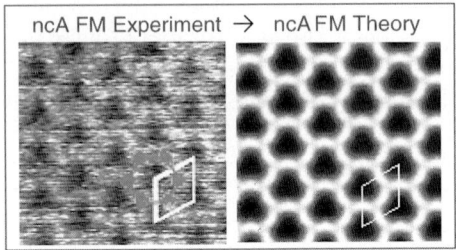

Fig. 23.2. nc-AFM image of Si(111)$\sqrt{3} \times \sqrt{3}$-Ag surface

and ncAFM images are quite different from each other, even though the sample surface is the same. On the other hand, theoretical simulation reproduces quite well both of the experimentally observed SPM images. It should be noted here that the bright spots in the STM image do not correspond to atoms, but the center parts of the triangles made up of three Ag atoms on the top layer. This assignment was crucial for the establishment of the HCT (honeycomb chained triangle) model of this surface [7]. On the other hand, theoretical simulation clarified that the network pattern appeared in the ncAFM image is caused by the structure fluctuation of the surface Ag atoms on the dynamic surface influenced by the nearby locating tip [8]. Moreover theoretical simulation has excellently reproduced the remarkable temperature dependence of the ncAFM image.

The above example demonstrates a crucial role of the theoretical simulation for understanding experimental results. However, the theoretical calculations so far tried have been very demanding, and one of the aim of this project is the development of much more efficient theoretical methods for the simulation. Toward this purpose the simplified and efficient calculation program, as well as user friendly graphical user interface (GUI) tools for the general users are aimed in this JST project. On the other hand, the other targets of this project are related to the development of a new frontier of the SPM simulation, in particular, for the experiments on the biological and organic molecular systems, and for the measurement in liquids.

23.3.1 Hierarchical Tip and Sample Model

For a realistic simulation of SPM, in particular AFM, not only the short range chemical interaction force between the tip and the sample, but also the long reaching interaction force ranging from meso to macro scale should be included at the same time. For the description of such multi-scale features, a hierarchical tip and sample model is now being developed in the JST project. In this model, the tip apex is modeled by atomic clusters made of several tens to several hundreds of atoms, and the tip basal part is modeled by an elastic continuum body. The elastic continuum is analysed by the finite element method, counting the force from the apex atomic cluster part together in a self-consistent manner. The same method is applied for the sample around the proximity region of the tip.

For the calculation of the force in atomistic part, phenomenological potential model such as Stillinger–Weber (SW) potential, MM3 potential, or embedded atom method (EAM) are used. The tight binding method based on density functional theory (DFT), so-called DFTB method, is also developed for the calculation of the short range force. The DFTB method makes it possible a simulation of the SPM experiments including chemical processes like the atom manipulation. This method is quite powerful and practical, thus it can be used for a variety of experimental situation in a flexible way.

23.3.2 STM Simulation for the Decorated Tip

A specific interaction between the tip and the molecular sample is very important for biological and organic SPM studies, and for this purpose the tip is decorated by the special molecular functional group. The SPM utilizing such decorated tip provides high sensitivity between specific functional groups, and detailed information is derived for the interaction force among macro-molecules or proteins. The JST simulator project is now developing the method of STM/AFM simulation using such modified tips with functional groups.

As for STM, the theoretical description of the coherent tunneling flowing through a molecule can be made possible by the non-equilibrium Green's function method. The same method, somewhat extended, can be applied for the current through chemisorbed molecule and modified tip. Introducing the interaction between the tip decorating functional group and the sample molecule on the substrate, the off-diagonal matrix element of the Green's function connecting the sample molecule and the functional group can be expressed using the Green's functions for the unperturbed system. Utilizing this relation, the tunnel current is obtained by the transmission probability of the electron from the tip side to the sample side. The final formula can be regarded as a slight extension of the formula for the STM current by the Bardeen's perturbation theory [1].

23.3.3 Theoretical Simulation Method for the Dynamic AFM in Liquid

As already noted, the harmonic oscillator model obtained by the projection to a particular normal mode of the motion of elastic continuum body of cantilever is particularly important for the analyses of the dynamic AFM in liquid. In this simulation method, we first solve the cantilever motion in liquid by the Navier–Stokes theory of the fluid dynamics. In the cantilever dynamics in the liquid, the nonlinear term in velocity of the Navier–Stokes equation is negligible, and thus the pressure to the cantilever surface is in proportion to the velocity and the acceleration of the cantilever at its local cross section vertical to the axis. Using this property, it can be shown that in the forced harmonic model in liquid, because of the pressure which is the counter action to the cantilever movement, the mass parameter and the friction coefficient are renormalized. By the preliminary calculations we found this effect is remarkably large and roughly explains experimental situation. The theoretical analyses of the cantilever oscillation is useful for the design of the ideal cantilever shape and the movement realizing a sensitive dynamic AFM. Furthermore the effect of the thermal fluctuation and inter mode coupling should be taken into account. The tip–sample interaction force mediated by the intervening water molecules is another important factor for performing reliable AFM simulation in liquid. Basic studies on this problem is now going on.

23.3.4 SPM Simulation for Organic and Protein Molecules

By using DFTB calculation, the adsorption structure of organic molecules for SAM membrane is clarified, and further, analysing the mechanical deformation of molecules by the tip of SPM, its influence on the STM or AFM images are revealed. As for the theoretical simulation of the AFM for the protein molecules, MD simulations and molecular relaxation methods are tried for simulating the SPM for systems including more than several tens of thousands of atoms. An example of such simulations, constant force AFM image of collagen molecules is shown in Fig. 23.3b. As the tip model for this simulation, a narrow carbon nanotube is used as shown in Fig. 23.3a. On the other hand, the effect of water molecules on the tip–sample interaction force immersing the collagen has been studied. The simulated image reproduces the corrugation pattern due to the helical structure of the collagen. It has been also clarified how the dynamical layer structure of the water is influenced by the presence of the collagen molecules.

To analyse larger molecules like proteins, the atomistic resolution may not be expected for many of the cases. Even in such cases, remarkable images and mechanical properties of the bio-molecules are observed and the quantitative analyses of these images are required. Moreover effects of mechanical response on various physical and chemical properties of proteins are attracting recent attention. For the theoretical analyses of these properties, a continuous elastic

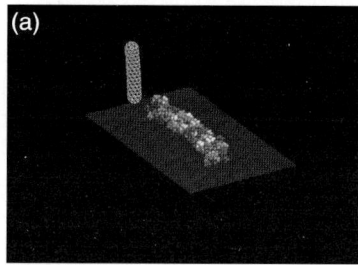 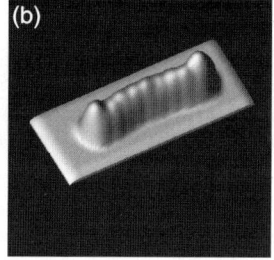

Fig. 23.3. The model of collagen and CNT tip: (**a**) simulation of the constant height AFM image of this system and (**b**) [9]

or viscoelastic model of the sample–tip system is often very plausible, and we study such method in the simulation of nano-mechanical properties of protein in the JST SPM simulator project.

23.3.5 The KFM Simulator

Kelvin probe microscopy can in principle afford a powerful experimental tool for measuring rich information of nano-materials such as local contact potential difference, polarization distribution, local dielectric function in atomic scale. However, methods for theoretical analyses of such quantities from the experimental data have not been developed so far. Recently, when the Si tip is approached closely to the Si surface, a remarkable structure has been found for the tip–sample interaction force spectra as the function of the bias voltage. Mechanism of such novel properties of the force seems to be related to resonantly enhanced force between the particular surface states of the tip and the surface. This feature should be solved by theoretical calculations. Generally tip–sample interaction force under a finite bias voltage is contributed by microscopic as well as macroscopic charge and capacitance distribution near the proximity region and the far field environment. Here again we encounter the problem of the hierarchical structure of the electric field and material response. The theoretical method mentioned in Sect. 23.3.1 would be extended for the quantum system including charge and the potential. The theoretical framework for this is now studied as the topics of the JST SPM simulator project. We also study a first-principles calculation method by the DFT for the open non-equilibrium system to be applied for the atomistic KFM simulation.

23.4 Present Status and Problems with Theories and Simulations of NSOM

Near-field scanning optical microscopy (NSOM) enables the observation of surface material distribution beyond the diffraction limit depending on the wavelength, and this observation can be interpreted in terms of two phenomena: (1) By illuminating an observation sample with evanescent light, a mutual interaction is generated between the optical probe and non-radiative near-field light which is localized within a smaller length from the sample than the wavelength of the light; (2) The electromagnetic wave produced by the mutual interaction is incident on the optical probe and the transmitted light passes into the probe and is received finally by a photo detector. It is therefore necessary to solve the 3D Maxwell equations for the multiple scattering due to the mutual interaction with the electromagnetic field.

In multiple scattering calculations for NSOM, self-consistent calculations have been done and a typical method is to find a self-consistent solution using Green's dyadic function, treating the optical probe as an ensemble of dipoles and the observed sample as another ensemble of dipoles [10, 11]. This method has the advantage of dealing with the range from the atomic level to the mesoscopic region. However, the actual calculation is not analytical; calculations for nano-sized objects require numerical calculations that use large amounts of RAM and involve huge computational numbers.

Instead of this method, the multiple–multipole (MMP) method using the macroscopic refractive index or the dielectric constant has been used as a global calculation method. In MMP, each domain is described by a series expansion of a known analytical solution that satisfies Maxwell's equations, and the unknown coefficients are determined to decrease the least-squares error between each domain wave at each boundary. This method has been used to do the calculations for a 3D probe [12]. However, problems with this method are that a suitable choice of known analytical solutions is difficult in general, and treatment of inhomogeneous or non-isotropic media is difficult. Since the electromagnetic wave boundary value problem is ascribed to a volume integral equation by using Green's function, a volume integral equation has been solved numerically in which the system of simultaneous linear equations is solved with an iteration technique [13]. But in the actual numerical calculation, the optical probe model is restricted to a sphere, and the material of both the probe and the observed sample forms the dielectric medium. In these numerical methods and the finite element method (FEM), a large matrix calculation is needed to calculate 3D nano-sized objects, and therefore a large amount of RAM and a huge amount of computation time is required.

The finite-difference time-domain (FDTD) method has been used not only in radio frequency applications, but also in optics applications and with NSOM [14]. FDTD is an intuitive method, because FDTD treats Maxwell's equations as finite difference equations in both space and time, and the wave calculation in all the cells is repeated from the initial value in time until a stable solution is obtained. Hence, this method has the following advantages: (1) No matrix calculation is necessary, and therefore a much smaller RAM capacity is required compared with other numerical methods. (2) Versatile modeling is possible, because the material in all the cells can be specified. (3) The wide frequency response of the optical system can be obtained in a single simulation. On the other hand, even this FDTD method used to have such problems as (1) Slow calculation speed, and (2) Need of large memory. But these problems accompanying the use of PCs have been solved with the recent rapid bump-up in speed and cluster techniques for PCs. Nowadays, using a simulator equipped with a GUI, modeling is easy, even in the case of a probe and a sample with a complicated shape made from a complex material. In addition, the problem of the unstable solution using a metal model with a negative dielectric constant, which is important in plasmon-enhancement applications, has been resolved. Using licensed commercial FDTD software equipped with these functions, numerical accuracy is confirmed by making calculations for analytical metal models with a negative dielectric constant, and a precise near-field optical design using this simulator has been done [15]. Moreover, the experimental results for an optical beam spot profile is in good agreement with the calculated results [16]. The thermal calculation for NSOM due to the heat generated by light absorption is well suited to FDTD, because the finite difference computational method can similarly be used for the thermal calculation, and an actual thermal simulation in cooperation with FDTD has been done. For more precise calculation, optical and thermal numerical computations taking into account both changes in the refractive indices and heat deformations of the probe and the sample should be performed.

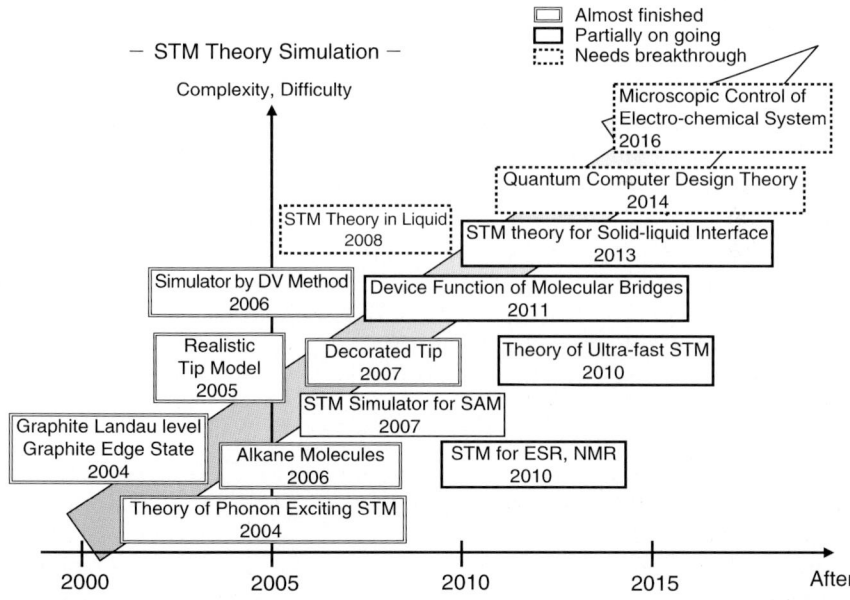

Fig. 23.4. Roadmap of STM theory

On the other hand, physical intuition and fundamental principles of mutual interaction cannot sufficiently be grasped with numerical calculation alone, so further analytical progress is desirable in this area. For example, approaches using diffraction theory or angular-spectrum representation [17] have been proposed.

So far, the explanation is based on classical electromagnetic theory. Since the mutual interaction between an optical probe and atoms or molecules is important in fields such as quantum dot spectroscopy and atom manipulation, the photon interaction effect based on quantum-mechanical theory has been calculated [18] and further progress is expected. Moreover, the combination of a global calculation and localized mutual interaction analysis will be needed in the future. Also, a breakthrough in near-field optics can be expected by calculation results by the unification of classical electromagnetic theory and quantum theory.

23.5 Roadmap

Road map of the development of the theoretical simulation studies and computational methods on STM, AFM/KFM and NSOM, are shown in Figs. 23.4–23.6, respectively. After the end of the JST SPM simulator project in a couple of years, researches on this field will be farther accelerated. The

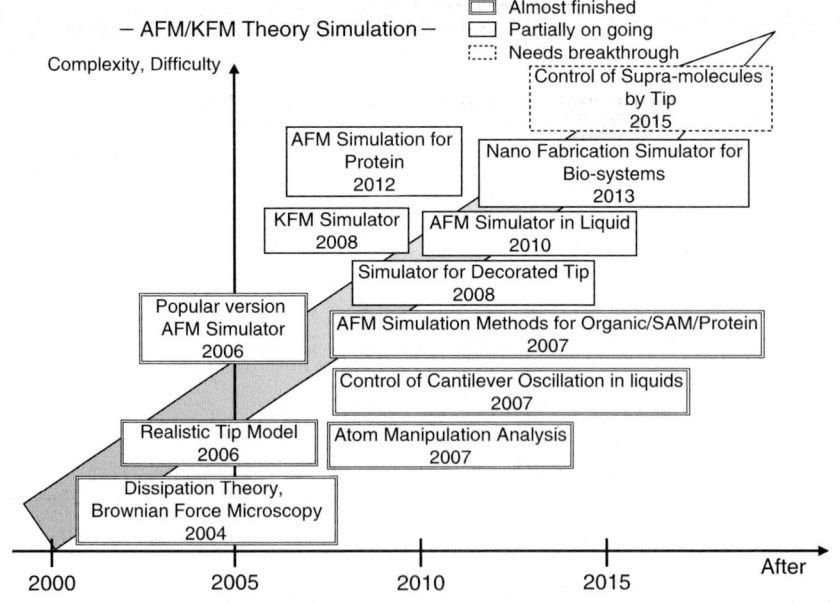

Fig. 23.5. Roadmap of AFM/KFM theory

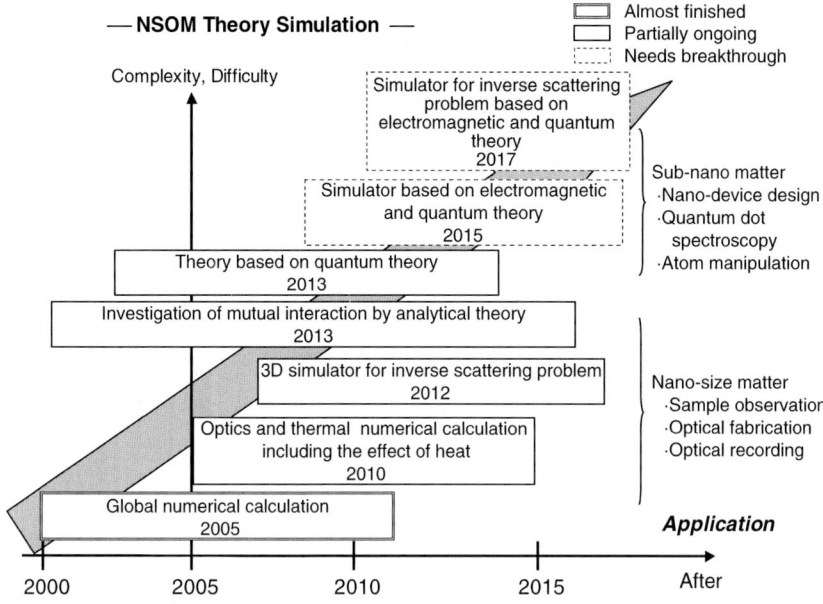

— NSOM Theory Simulation —

Fig. 23.6. Roadmap of NSOM theory and simulation

theoretical analyses of STM observations of SAM membrane and organic molecules chemisorbed on surfaces should be remarkably developed along with the development of molecular electronics. In particular the simulation of SPM with decorated tip with functional group is important. How is the tunnel current through molecular bridges connecting between the electrodes governed by the linkage structure between the electrode and the molecule, is the basic issue in the molecular electronics. Theoretical studies will give an important clue for this problem. Unresolved theoretical problems in the theory of STM is the simulation method for the measurement in liquids, as well as for the electrode processes of the electrochemical cell. These simulations are particularly important for the measurement of biological samples, and challenges from the theory are needed. The method to treat the water mediated interaction between the tip and sample is urgently needed for that purpose.

The development of an efficient AFM simulation for organic molecules and proteins will stimulate the experiments on quantitative nano-mechanics of biological macro-molecules. The research on 3D dynamic structures and intermolecular interactions will be greatly enhanced with the development of theoretical approaches.

For theoretical research into NSOM, further development in numerical calculation and mutual interaction analysis will be needed. In particular, unification of classical and quantum theory will be necessary in the future as well as the development of simulators to solve the inverse scattering problem in which the true optical image is calculated from the experimental image.

References

1. M. Tsukada, K. Kobayashi, N. Isseki et al, Surf. Sci. Rep. **33**, 265 (1991)
2. T. Sasaki, U. Nagashima, M. Tsukada, J. Chem. Softw. **6**, 165 (2000) (in Japanese)
3. M. Tsukada, K. Tagami, K. Hirose et al, J. Phys. Soc. Jpn. **74**, 1079 (2005)
4. T. Mii, S. Tikhodeev, H. Ueba, Surf. Sci. **502–503**, 26 (2002)
5. T. Schimizu, K. Kobayashi, M. Tsukada, Appl. Surf. Sci. **60/61**, 454 (1992)
6. M. Tsukada, N. Sasaki, K. Tagami, Solid State Phys. **38**, 257 (2003)
7. S. Watanabe, M. Aono, M. Tsukada, Phys. Rev. B **44**, 8330 (1991)
8. N. Sasaki, S. Watanabe, M. Tsukada, Phys. Rev. Lett. **88**, 0461061 (2002)
9. K. Tagami, M. Tsukada, J. Surf. Sci. Nanotechnol. **4**, 294 (2006)
10. C. Girard, D. Courjon, Phys. Rev. B **42**, 9340 (1990)
11. K. Kobayashi, O. Watanuki, J. Vac. Sci. Technol. B **14**, 804 (1996)
12. L. Novotny, D. W. Pohl, B. Hecht, Ultramicroscopy **61**, 1 (1995)
13. K. Tanaka, M. Yan, M. Tanaka, Opt. Rev **9**, 213 (2002)
14. H. Furukawa, S. Kawata, Opt. Commun. **132**, 170 (1996)
15. S. Hasegawa, F. Tawa, Appl. Opt. **43**, 3085 (2004)
16. W. Nomura, M. Ohtsu, T. Yatsui, Appl. Phys. Lett. **86**, 181108 (2005)
17. T. Inoue, H. Hori, Opt. Rev. **5**, 295 (1998)
18. K. Kobayashi, S. Sangu, M. Ohtsu, *Progress in Nano-Electro-Optics I* ed. by M. Ohtsu (Springer, Berlin Heidelberg New York, 2002) pp. 119–157

When Will SPM Realize Our Dreams?
The Roadmap of SPM

Osamu Kubo

In this chapter, the discussion on the following themes are summarized: (1) Is there any substitute for semiconductor devices? (2) Future scientific and/or social prediction from the viewpoint of the demands. (3) The orientation of SPM: Future prediction from the viewpoint of SPM technologies.

(1) Is There Any Substitute for Semiconductor Devices?

Morita: As you all know, conventional LSI circuits are rapidly miniaturized, and, by the year 2018, it is said that the width of the interconnects would be as small as 10 nm, so that the processing accuracy would be 1 nm, and measuring accuracy would still be its tenth, 0.1 nm, almost to an atomic level. At first, we would like to discuss if there is any technology to take place for the current fine processing technology of semiconductors, and a microassembly such as bottom-up method be of any use. In fact, there still are no reports on technology corresponding to a process in microassembly, on a system of elemental devices and on integration for cost reduction. There is no possibility that microassembly replaces a fine processing technology without progress of above-mentioned issues. It is getting too late to take place of the fine processing technology if there is no prototype of a process based on microassembly.

Hosaka: The electric current had been controlled by the potential in the vacuum tubes that were previously prevailing. On the other hand, transistors are switched with changing resistance by electric field. So, these two have a slight difference in the principle of operation. What might be used for the device formed by microassembly of atoms and/or molecules?

Morita: Radios, televisions, and computers have been made both with transistors and vacuum tubes. But if it comes to making them by using atoms and/or molecules, it is not totally in vision; there is actually no vision of elemental device and its manufacturing. At present, the microassembly at an atomic level is only discussed in biotechnology. Only 30,000 bases function as gene in a DNA of living cell, but that 30,000 bases control the 60 trillion cells of human body, which means that it has compressed so much of the data. In the point of speed and accuracy, semiconductor devices are very quick and rigid systems, while living organisms can be compared to a loose system, as its response speed is in a range of millisecond and mutation and evolution occurs at times. There has been mention of brain computers, but there are some uncertainties, such as the difference with current computers. It still remains a question if it would be able to produce something like television and/or computer from living organisms.

Yamada: One direction of studies on molecular electronics is in an extension of the current transistor development, i.e., to realize the same thing in molecules as done in silicon devices made by top-down methods. Living organisms succeed in efficient information processing using minimal energy. While it is ideal to use such features of biomolecules for the development of devices in molecular electronics, a present state is that researchers are making something close to silicon devices because the specific direction of studies is not yet clear. Many researchers hope eventually to adopt the features of living organisms such as self-organization, self-repairing, and autonomous distribution. It would be amazing if molecules can make transistors and fabricate devices by themselves.

Nakamura: It is a common recognition among researchers using organic materials that "replacing silicon devices" is what they never should mention. For the people using solid-state devices, there is no necessity of replacing the most current silicon devices, nor any other candidate. If we consider which part of atomic and/or molecular devices can take place, we must assume the year beyond 2018, which is said to be the limit of development of current semiconductor devices in the point of fine processes. On the other hand, there is a trend of using macroscopic electronics to what has not been used and with less cost; for instance, a carpet having judgment ability. Even if the scale is large, atoms and/or molecules do provide functions, and many things happening are found for the first time when seen microscopically. It is not necessary to make a device and/or system in nanoscale just because you are observing a nanoscale substance. We do not have to decrease the area of activity. The brain operates only as a group of cells, although it is unknown how each cell works. To understand a group in a scientific approach, we must understand the microscopic elements and then the macroscopic systems. Even if one handles macroscopic systems, our tools, nanoscale evaluation techniques, will be extremely important to understand the elements of the systems.

Morita: There is a big difference between current electronics and living organisms. In the electronics, electrons are moved by charging energy from outside, so that the heat would be the cause of noises. On the other hand, living organisms utilize that heat energy. If we would be able to make and move something using heat energy, something totally different might emerge. Although silicon is a single crystal made of a single element, multi-element system would be one of the keywords from now on. The reason we have used single crystal made of a single element is to control currents by adding just a little impurity to a refined electronic band structure. It is time to think of fabricating devices and systems with a complex or multi-element system that would completely change the past studies.

(2) Future Scientific and/or Social Prediction from the Viewpoint of Demands

Nakajima: People studying organic materials or biomaterials are seeking for a new computing. In that case, sizes and speeds are not the matter. What is more important is whether they could solve problems that cannot be solved with algorithm nowadays or not. It is not practical, for example, to take a million years to solve a problem with a binary computing even if the problem is not substantially unsolvable. The new algorithm might enable us to calculate such problems in a practical time. The problem is that the correct understanding of such algorithm has not begun yet. It is desirable to have an observing point as what is happening in biomaterials and/or organic molecules; such observations are needed.

Katayama: As Dr. Nakajima said, computers have not at all progressed in terms of algorithm or architecture. If the understanding of brain algorithms realizes, we should make a new computing system, at least on the basis of silicon devices, using current fine processing technology. It is not necessary to use molecules if we are oriented to changing the architecture from the current technology.

Tanaka: I am pretty much aware that the architecture is not progressing well. Considering the structure of one element of a transistor, are the devices still in a progress? The basic structure is defined, and it seems emphasized on how it would be made smaller, which also would need some break-through.

Katayama: If we ask the users whether improvement of computers make their livings even better, the answer is probably no, they are already satisfied. What we need are purposes, demands, and contents that are, for example, environment-friendly, to meet aging society, and so on. It is nowadays nonsense that data as large as that of the National Diet Library could be gathered to something small as a sugar cube, which Pres. Clinton formerly mentioned. We now can gather any data from any place through the infrastructures, which also must be taken into consideration.

Morita: In that point, the information technology has already come to a certain level, and it is time that the overall social system and frameworks have almost come to a change through the networks.

Tomitori: When something different from the conventional man–machine interface appears, a new frontier might come to an open. Computers take the speed and the exactness role, and the point is how they are made accessible to human brains. For instance, the technology of attaching a sensor-like device to a person enables us to know his/her physical condition. If such technology evolves, in other words, if computers are enforced to make joint with the brain, a completely different computing system might appear. Some day nerves and neural networks may be connected directly to silicon computers.

Furukawa: Say, for instance, we cannot mechanically attach an arm to a person who lost his/her arm in a car accident. Such interface has not yet completed. Molecular devices may play a role in such situation. In that case all the functions including those of brains do not have to be concerned of. Semiconductors also must have begun with no idea what would it be used for. But when transistor radios appeared, everyone soon came up with ideas of what they could do with. Molecular devices also might surprisingly make progress if something like transistor radio in semiconductors would be realized.

Kadota: People may not require standardized products but something that has brain or evolves to some extent. The performance of semiconductor devices itself is improving, but it does not always satisfy the users. There are many unnecessary functions.

Nakamura: It may not be so easy to change hardware devices unlike software upgrade through the web. However, a nanoscale device would not need so much energy to be rearranged (upgraded).

(3) The Orientation of SPM: Future Prediction from the Viewpoint of SPM Technologies

Hasegawa: What would SPM aim at, the precision of transistors, or a different direction such as functions of living organisms even if the precision is sacrificed? Would it be needed to reconsider how to use SPM?

Morita: You mean whether SPM is enough only to be used for measurements or analysis. It is time we need to consider the essential aim of the bottom-up technologies like microassembly, not only as the fusion of them with the current top-down processes. The original orientation of STM was making towards better precision, and has come up to analyzing in a highly extreme condition of low temperatures and/or high vacuums. From now on, I feel it is time to search for more new orientations with using at room temperature, in solutions and in air, not simply for the analysis.

Katayama: Twenty-five years since STM came up, there have been many findings. But fundamentally no new principle or physical phenomenon has been found. It is necessary for nanoscience to find some new physical phenomenon and certain principle to open a different paradigm, but such basic science has not yet been achieved enough.

Tomitori: On the other hand, because quantum physics is very well established and completed, it explains most of the phenomena in nanosize. In that term we ought to stay focus on the applications rather than pure science to spread our view, because it may be difficult to transcend quantum physics.

Morita: Quantum physics exists as physics of extremely microscopic world, so we must think of another way. One direction is to think complex and systematic devices as in semiconductors. In biology, DNA, proteins, and cells emerge different functions from the isolated ones only after they form into a large group. We study on single atoms and electrons using SPM, however, the difference might not emerge unless we handle the enlarged size.

Nishikawa: Viruses and living organisms are complexes, but where does the biologic function come from? Tobacco mosaic virus, for example, crystallizes and also grows itself but it is not a living organism. Even though the sequence of atoms/molecules is elucidated, what gives life is unknown. This may refer to the problem of what is the border of life and death, which might be clarified if life could be made by assembly techniques.

Morita: I also do not know what life is. As for myself, however, I wish to make something that works on the basis of a completely different principle.

Yamada: One of the further extensions of SPM is to develop as an advanced analysis and/or evaluation method, and is a realistic orientation in the multifunctional analysis. Another possibility is SPM technology for manipulation, assembly and control of atoms/molecules. It would be a key point whether such technologies are applied to the actual use in manufacturing processes. Referring to molecular devices, is it able to make something smaller and smaller with the combination of SPM and a method based on printing technologies? Such application, as dip pen lithography, is gradually spreading, but not yet among the people originally using SPM.

Morita: There are people developing "Millipede". But SPM is rather a tool for searching or for the initial stage of assembly, and it may be practical to cut-off and to find another way for the mass production. Taking into account control of them, it might be tough.

Yamada: Recently, many people studying MEMS are joining the SPM-relating researches, which bring possibility that the progress of cantilevers would extend to practical manipulation and/or assembly techniques in the future. In the present state, the specific studying course is not clear, but we hopefully would want to somehow apply the SPM technology to products in the near future.

Morita: There are examples of good outcomes in an area no one had studied, a few of which introduced in this book. These are the examples of the results of making steady efforts that was said to be high-risk-high-return and to take another 5–10 years, and have succeeded in making a breakthrough. Our area remains a real breakthrough if concentrated on the right study.

Nakamura: What I thought about the studies concerning "atom inlays" of Morita lab. (see Chapter 11.2) was that it would be similar to a high-temperature, high-pressure condition under the probes. It is beneficial to make use of such condition for processing.

Furukawa: It would be quite interesting, for example, if we can make a new composition of materials even in a very limited scale, measure its property, and find its new feature. Arranging atoms has been technically achieved. If it would be possible to measure electric and physical properties of such arrangements, it could give an impact on extensive areas.

Katayama: We are working on making FET and logic circuits with nanowires and nanotubes as nanodevices, and the surrounding interconnects are totally formed by top-down methods. If we understand the function of such devices by probes without interconnects, using for example double probe SPM, it will be great in the point that we could understand the function right after the formation. It would be necessary not only to use spectroscopic measurements but also to actually flow current through nanostructures when we want to search for their new functions.

Morita: As long as we consider electronic devices with something like atomic wires, there is no way than flowing current. In this case, contacts between electrodes and devices have to be made highly reproducible. Our idea is to make in atomic scale not only the wires but also the parts of initial contact, which are connected to macroscopic electrodes by top-down method. In this way the contacting parts would scarcely affect the property of the device, which enables us to search the devices to some extent.

Nakamura: As concerning with observation tool, the study of three-dimensional tomography with EFM has been stopped (see Chapter 6.2). I would want to take the three-dimensional tomography, because electrostatic force is a comparatively long-range force, and might enable us to see the inside of materials. If we can see where the charge is carried inside a biomolecule which has certain thickness and function within, there may be a new discovery.

Morita: An SPM has a restriction that it can only see the surface of the object. If three-dimensional analysis as in computer tomography can be made, however, an extremely huge and new possibility will be made. It may be difficult to reach as far as atomic resolution. However, long-range force should, in principle, contain information of deeper area, so it would be possible to derive data from as deep as to several surface layers by making a program that can distinguish the data of each layer. There is a possibility in the way that we make a standard sample and develop the matched programs to it.

. . . .

Although there are further discussions, I would like to conclude this paper discussion. I sincerely wish the above-mentioned dreams would soon be realized.

Index

Printing: Krips bv, Meppel
Binding: Stürtz, Würzburg